D1339944

LIVING FREE
The Story of Elsa and her Cubs

Joy Adamson was born in Austria in 1910, and moved to Kenya in 1938. Here she made a name for herself as a painter of flowers and plants, and was commissioned by the government to make a series of paintings to record the Kenya tribesmen. In 1943 she married George Adamson, then a Game Warden. It was as a writer that she was to achieve worldwide recognition and personal fulfilment, beginning with the three books in which she described the raising of the lioness Elsa and her cubs and how she re-adapted them to the wild, *Born Free* (1960), *Living Free* (1961) and *Forever Free* (1962). In time the Adamsons went on to raise and successfully return to the wild two more of Africa's great cats: in addition to Elsa there was the cheetah, Pippa, the subject of *The Spotted Sphinx* (1969) and *Pippa's Challenge* (1972), and a leopard called Penny, described in *The Queen of Shaba* (1980). In 1980 Joy Adamson died, the victim of a brutal murder – and nine years later her husband was also tragically to be murdered in the bush.

The Adamsons' tireless work on behalf of Africa's endangered wildlife continues in the charity they established, The Elsa Wild Animal Appeal.

By the same author

Born Free
Forever Free
The Peoples of Kenya
The Spotted Sphinx
Pippa's Challenge
Joy Adamson's Africa
The Searching Spirit
Friends from the Forest
The Queen of Shaba

Joy Adamson

LIVING FREE

The Story of Elsa and her Cubs

Introduction by
Sir Julian Huxley FRS

Fontana/Collins Harvill

The publishers gratefully acknowledge the assistance of
ICCE, the International Centre for Conservation Education,
in supplying prints of the photographs.

First published by Collins and Harvill Press 1961
First issued in Fontana Paperbacks 1964
8 Grafton Street, London W1X 3LA
This edition with new cover issued 1990

Phototypeset by Input Typesetting Ltd, London

Map by M.L. Design

Printed and bound in Great Britain by
William Collins Sons & Co. Ltd, Glasgow

To all wild animals
and their freedom

AUTHOR'S NOTE

Since Elsa's cubs were born I have kept notes in which I have recorded what we observed of Elsa and her family when we were in camp. *Living Free* is based on these notes and this explains its form, and the use of the present tense in relation to Elsa.

All the experiences recounted in this book were shared with my husband George, and it could not have been written without him. I would like to acknowledge my gratitude to everyone who, whether in an official or a private capacity, made it possible for us to spend so much time at Elsa's camp. I would also like to thank all those who helped me with *Born Free* and have helped again in the publication of *Living Free*.

CONTENTS

ILLUSTRATIONS

I came upon Elsa trotting along with the cubs
As the cubs would not cross the swollen river for fear of the crocodiles,
  Elsa swam across to them with a carcase
Family portrait
Elsa, Christmas 1960

INTRODUCTION

Last September my wife and I had an unforgettable experience. We saw Elsa, followed by her three cubs, burst into the clearing in the Kenya bush where the Adamsons periodically camped. The cubs sat themselves down to look and watch, interested but aloof, while Elsa sprang towards Joy Adamson as towards an intimate friend, putting her great paws on Joy's shoulders and almost knocking her over with the vigour of her greeting.

So it was really true . . . True that a full grown lioness, after she had established a strong emotional attachment to Mrs Adamson and her husband, had been deliberately left in the wild bush, had found a wild mate, had produced those wild-born cubs and yet had retained this personal involvement with her human friends.

You may quarrel with that word *personal* as applied to a mere animal. But, after having seen Elsa with the Adamsons as well as having read Mrs Adamson's two books, I insist that it is the right one. By a passionate patience and an understanding love, Joy Adamson succeeded in eliciting something in the nature of an organized personality out of an animal's individuality, set of its instincts strung on the simple thread of its memory.

Of course, something of the sort can happen with dogs or with chimpanzees. But in such cases the emergent animal personalities are elicited in domesticated or captive creatures, unable to escape from their captivity into full natural freedom. But Elsa was fully integrated with the life of the wild. She still sought out a wild mate when on heat, and kept up communication with him even when regularly visiting the Adamsons; she jealously guarded her newborn cubs against discovery by her human friends until such time as she deliberately brought them into camp; when the Adamsons were elsewhere, she killed her own prey; she received fearful wounds in fights with wild rival lionesses. And yet the human attachment stayed unimpaired.

I find this not only interesting but moving. The story of Elsa, set forth in detail in this and Mrs Adamson's previous book, demonstrates the wealth of potentialities in higher mammals, waiting to be drawn out and elicited into actuality. And it shows that the best and perhaps the only method of eliciting those hidden potentialities in any fullness is through emotional but intelligent involvement, by way of what I have called understanding love.

This, I think, is important. It is important for the progress of science. It means that in the young science of Animal Behaviour (or Ethology, as it is now called), the investigator will only obtain his most valuable results by supplementing his scientific objectivity with an understanding and even affectionate approach to the animals with which he is working. This applies with special force to any attempts to discover the extent of unrealized possibilities latent in his animal subjects. It is important for animal trainers and zoo keepers and officials: the porpoises at Marineland will not stand for a cross word, let alone punishment: the great apes have a deep need for some sort of personal relationship with their keepers, and even such an apparently lethargic character as the Giant Panda responds to an intimate approach. It is important for human education, as progressive educationalists have long discovered; and for all attempts at contact between hostile or mutually suspicious groups of people, as the modern world is beginning to find out. Patience and understanding, backed by friendliness and a spirit of love, could be as effective here as they were with Elsa.

But I must not spend more of my space on generalities. *Born Free* was the story of Elsa: *Living Free* is the story of Elsa and her three cubs, from the time of their birth to shortly before her untimely death. I found it an absorbing story. First, because it gives the reader the genuine feeling of the African bush. The nocturnal buffalo which knocked Mrs Adamson down and played havoc with the camp like the proverbial bull in his china shop. The aggressive ratel which could force the full-grown Elsa to retreat: the troops of elephants trumpeting and rumbling nearer and nearer in the dark. The gathered vultures watching for death, the Bush-babies peering out of the branches with their big round eyes, the swarm of newly-hatched crocodiles in the river; the chattering baboons, the guinea-fowl, the comic parrots; the beautiful little genet which came to eat the Adamsons' supplies; the hippos ponderously sporting in the pools, the startled buck crashing through the bush, the hyenas that tried to steal the Adamsons' meat; the distant roaring of lions in the night.

But that, fascinating though this is, is only a background for the story, only the stage on which the protagonists live out their parts. The main interest of the book lies in its account of the psychological development of Elsa and her family.

Again, I expect that purists will quarrel with the word *psychological: behavioural* is now the orthodox term. But again I insist that it is correct. We all, including even the most 'scientific' scientists, conduct our lives in the belief that other men have minds and emotions and are capable of subjective experience, even though we can only deduce this from their observable behaviour (including, of course, their vocalized behaviour or speech). And psychology would not be much of a science if it did not draw on subjective experience and failed to base itself on a belief in the importance of the mental aspect of our two-faced natures. Higher vertebrates, and especially the higher mammals, have brains and behaviours similar to ours in many essential ways. Why deny them psychological experiences similar to ours? Darwin did not do so. By the very title of the book which founded ethology as a science – *The Expression of the Emotions in Man and Animals* – he affirmed the similarity and rejected dogmatic Behaviourism.

When people like Mrs Adamson (or Darwin for that matter) interpret an animal's gestures and postures with the aid of psychological terms – anger or curiosity, affection or jealousy – the strict Behaviourist accuses them of anthropomorphism, of seeing a human mind at work within the animal's skin. This is not necessarily so. The true ethologist must be evolution-minded. After all, he is a mammal. To give the fullest possible interpretation of behaviour he must have recourse to a language that will apply to his fellow-mammals as well as to his fellow-man. And such a language must employ subjective as well as objective terminology – *fear* as well as *impulse to flee, curiosity* as well as *exploratory urge, maternal solicitude* in all its modulations in welcome addition to goodness knows what complication of behaviourist terminology.

This is not to deny that behaviourist methodology – the rigorous analysis of observable behaviour, if possible on a quantitative and experimental basis, without psychological assumptions or interpretations – is essential for the progress of ethology. Of course it is, as shown by the brilliant work of men like Tinbergen or Hinde. But by itself it will not give us a rounded interpretation of behaviour, a full comprehension of its biological significance as the outward and visible sign of mind in its upward evolutionary progress.

Furthermore, such analysis is not always possible. Physicists and chemists, and indeed laboratory scientists in general, tend to regard field study and observation as scientifically inferior. They forget that all branches of science, including their own, begin by being observational. They also forget the fact that physics and chemistry are far simpler subjects than biology, and that biological organization reaches patterns of complexity many thousand times greater than anything in the inorganic world.

This last point is especially relevant for ethology. For one thing, it has to rely on field observation to provide much of its subject-matter in the first instance. For another, captivity always imposes some limitations on an animal's actions, and experiment can never duplicate the range of situations that may confront an animal in nature. Accordingly, many of the most complex and subtle (and therefore the most interesting and significant) animal activities will not be manifested in the experimental laboratory. They must be searched for and observed in free nature. When found, they must be interpreted in the analogical light of the complexities and subtleties of our own mind-accompanied activities, and when so interpreted, they both enlarge and enrich the scope of science.

Mrs Adamson's observations on Elsa and her cubs have enlarged mammalian ethology in ways which would not have been possible with animals in captivity. We see Elsa after the birth of her cubs jealously guarding the secret of their whereabouts, deliberately misleading the Adamsons in their search; later when the cubs are no longer vulnerable, she as deliberately brings them into camp and displays them to her human friends. We see the three cubs developing three very different individual temperaments – from Little Elsa, so timid that she could never approach the human circle, to Jespah who roamed through the tents and nibbled at human toes. He was his mother's favourite and was jealous of anyone who got between him and his mother's affection.

We see the cubs at play, romping like kittens, playing with their mother's twitching tail, frolicking in the river, climbing trees as adventurously as schoolboys, and Elsa joining in their games. Elsa shows annoyance by putting her ears back and narrowing her eyes. She submits to the painful extraction of a thorn from her tail with the patience of Androcles's lion. She will have nothing to do with strange Africans, but treats white strangers – and also the Africans she knows – as friends. She distinguishes between the sound of a plane and of a car. She is annoyed by the wireless, but (like the cubs) pays no attention to a sudden burst of flame. She suffers greatly from tsetse flies, and comes to Mrs Adamson to help her be rid of them.

We see the cubs at fourteen weeks burying surplus meat; and at ten and a half months with incipient manes, but still suckling. We see Elsa, at a dead zebra provided for her by the Adamsons, standing and roaring loudly instead of falling to, apparently as a notification to her wild mate across the river. We are told of her remarkable behaviour to her publisher, Mr William Collins, when he came to visit her – how she twice broke into his tent, and once took his neck and cheeks between her great jaws (Mrs Adamson thinks that this was a sign of affection, but doesn't seem quite sure). I would add that I think Mr Collins's behaviour at the time was equally remarkable, and also his behaviour in returning later in the year to visit her again.

But most remarkable of all is the fact I started from – the fact that a human being had succeeded in eliciting in a lioness a psychological organization which basically resembled a human personality. And that in consequence the lioness was enabled to lead a second life, based on friendly human relations, in addition to a normal animal life in the wild.

All in all, *Living Free* is a remarkable story, as extraordinary as *Born Free*, and in many ways more interesting.

JULIAN HUXLEY
1961

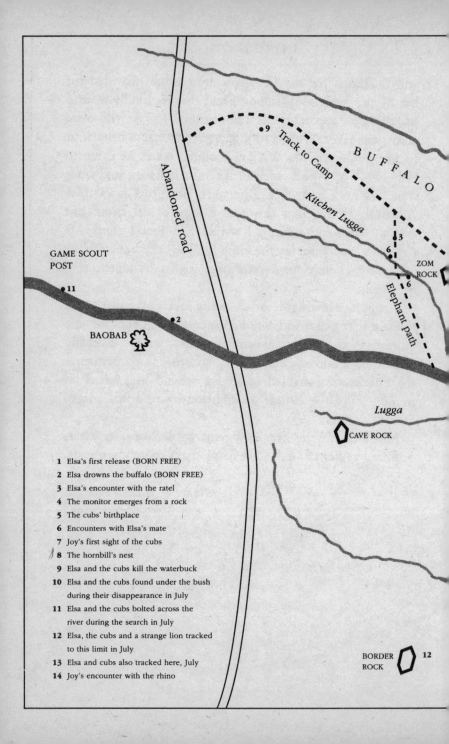

BUFFALO

●9 Track to Camp

Abandoned road

Kitchen Lugga

3

6   ZOM
     ROCK

6

GAME SCOUT
POST

Elephant Path

●11

●2

BAOBAB

Lugga

CAVE ROCK

1 Elsa's first release (BORN FREE)
2 Elsa drowns the buffalo (BORN FREE)
3 Elsa's encounter with the ratel
4 The monitor emerges from a rock
5 The cubs' birthplace
6 Encounters with Elsa's mate
7 Joy's first sight of the cubs
8 The hornbill's nest
9 Elsa and the cubs kill the waterbuck
10 Elsa and the cubs found under the bush
    during their disappearance in July
11 Elsa and the cubs bolted across the
    river during the search in July
12 Elsa, the cubs and a strange lion tracked
    to this limit in July
13 Elsa and cubs also tracked here, July
14 Joy's encounter with the rhino

BORDER   12
ROCK

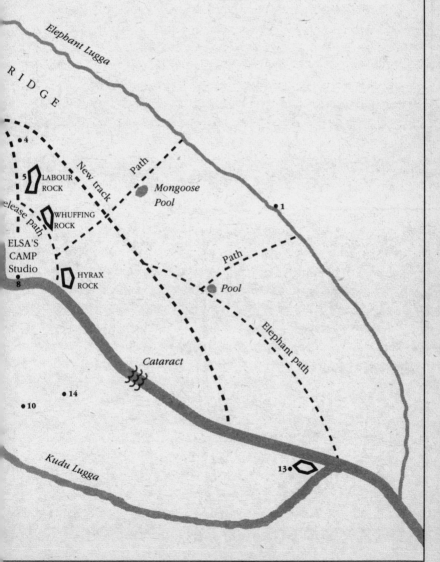

ELSA'S CAMP

0   1   2

MILES

Elephant Lugga

RIDGE

• 4

New track

Path

5 LABOUR
ROCK

*Mongoose
Pool*

• 1

Release path

WHUFFING
ROCK

ELSA'S
CAMP
Studio

HYRAX
ROCK

Path

• 8

Pool

Elephant path

Cataract

• 14

• 10

Kudu Lugga

13 •

1

Elsa Mates with a Wild Lion

It was between 29th August and 4th September 1959 that my husband George actually saw Elsa and her lion courting. Quickly he made a calculation – 108 days' gestation – this meant that cubs might arrive between the 15th and 21st December.

When on his return to Isiolo he told me what he had seen I could hardly bear not to start off for camp alone, for I was afraid that Elsa might now follow her mate into a world beyond our reach.

But when we arrived she was there waiting for us by the big rock close to the car track.

She was very affectionate and also very hungry.

As our tents were being pitched her lion started calling and during the night he circled round the camp, while she remained with George eating heartily and quite uninterested in her mate's appeal. At dawn we heard the lion still calling but from much further away.

For two days she remained in camp eating so enormously that she was too sleepy to move till the afternoon when she went fishing with George.

During the third night she ate so much that we were quite worried about her; yet in the morning, in spite of her bulging belly, she trotted into the bush with us and first stalked two jackals and then a flock of guinea fowl.

Of course, each time she closed in on them they flew off, whereupon she sat down and licked her paws. I was walking ahead but stopped dead at the sight of a ratel; this animal, also known as a honey badger, is rarely seen. It had its back turned towards me and was so absorbed digging for grubs in the rotten wood of a fallen tree that it was quite unaware of Elsa's approach. She saw it and crept forward cautiously till she was practically on top of it.

Only when their heads nearly bumped together did the ratel take in the situation; then hissing and scratching he attacked her with such courage and so savagely that she retreated.

Using every advantage that the ground offered the ratel made a fighting retreat, charging often, and eventually disappeared none the worse for his adventure.

Elsa returned defeated and rather bewildered; plainly she was too well fed to hunt except for sport and there was no fun to be had with such a raging playmate.

This incident made us sure that we had been right in suspecting a ratel when, in the early days of Elsa's release, we had found deep bites and gashes on the lower part of her body. For no other small animal is so fearless and bold.

On our walk home Elsa, full of high spirits and affection, rolled me over several times in the sand, while I listened to the trumpeting of elephants which were much too close for my liking.

That night she slept in front of my tent, but just before dawn her lion started calling and she went off in his direction.

Their calls were easy to distinguish; Elsa has a very deep

guttural voice, but after her initial roar only gives two or three whuffing grunts, whereas her lion's voice is less deep and after his roar he always gives at least ten or twelve grunts.

During Elsa's absence we broke camp and left for Isiolo hoping that she was in the company of her mate.

While I was at Isiolo George was asked to look after a baby elephant which had fallen into a well. Of course he brought it home. We called it Pampo; it was a most engaging creature and well worth the trouble it caused us, which included supplying it with two gallons of milk a day.

Baby elephants abandoned by their mothers can rarely be reared for it is very difficult to find a substitute for an elephant's milk which has a different content from any other, so although we added cod liver oil and glucose to Pampo's diet, I felt very anxious about his future and scarcely left him for an hour.

Housekeeping for two animals who lived one hundred and fifty miles apart was a problem. We couldn't neglect Elsa, nor allow Pampo to die for want of care. Luckily, my friend Joan Jugl, who is a great animal lover and has had much experience in handling them, offered to act as elephant sitter, so we were able to return to the camp on 10th October.

It was three weeks since we had left Elsa; an hour after our arrival we saw her swimming across the river to greet us, but instead of the exuberant welcome she usually gave us, she walked slowly up to me. She did not seem to be hungry and was exceptionally gentle and quiet.

Patting her, I noticed that her skin had become extremely soft and her coat unusually glossy. I saw, too, that four of her five nipples were very large.

She was pregnant. There was no doubt about it. She must have conceived a month ago.

It is widely believed that a pregnant lioness who is handicapped in hunting by her condition, is helped by one or two other lionesses who act as 'aunts.' They are also supposed to assist in looking after the new-born cubs, for the male is not of much practical use on such occasions and, indeed, is often not allowed near the young lions for some weeks.

Since poor Elsa had no 'aunts,' it would be our job to replace them. George and I talked over plans to help to feed her and avoid any risk of her injuring herself during her pregnancy.

I was to stay in camp as much as I could and, at the nearest Game Scout Post, some twenty-five miles away, we would establish a herd of goats from which I could collect a few in my truck at regular intervals.

Nuru would remain with me to help with Elsa and Makedde would guard us with his rifle, Ibrahim could drive and I would keep one boy, the Toto (the word Toto means child in Swahili), to act as a personal servant.

George would visit us as often as his work allowed.

As though she had understood our conversation, Elsa hopped on to my camp bed as soon as it was made ready and looked as if she thought it the only suitable place for someone in her condition.

From now on she took possession of it, and when next morning, as I did not feel well, I had it carried down to the studio, she came to share it with me. This was uncomfortable, so after a time I tipped it over and rolled her off. This indignity caused her to retire, offended, into the river reeds till the late afternoon when it was time for our walk.

When I called her she stared at me intently, advanced determinedly up to my bed, stepped on to it, squatted, lifted her tail and did something she had never before done in so unsuitable a place.

Then with a very self-satisfied expression she jumped down and took the lead on our walk.

Apparently, now that she had had her revenge everything was again all right between us.

I observed that her movements were very slow and that even the noise of elephants close by only made her cock her ears. That night she rested in George's tent, unresponsive to the call of a lion who seemed to be very near the camp.

As in the early morning the lion was still calling, we took Elsa for a walk in his direction. There, to our surprise, we found the spoor of two lions.

When she began to show an interest in these pug marks we left her and returned home. She did not come back that night, so we were surprised to hear a lion grunting extremely close to the camp. (Indeed, in the morning his pug marks proved that he had been within ten yards of our tent.) The next day Elsa again stayed away. Hoping to make the lions kindly disposed towards her, George shot a buck and left it as a farewell gift; then we returned to Isiolo.

We were delighted to find Pampo well, though unfortunately he was beginning to share the fate of all celebrities and attracted a host of visitors. I was worried about this because young animals are usually very sensitive to the presence of strangers and I had reason to think that his admirers tired him and made him nervous for as soon as he and I were left alone he would trustfully move his bulky body against mine and then go to sleep. Plainly this contact gave him a sense of security.

After we had spent two weeks at home we decided that it was time to go and see Elsa. Joan Jugl very kindly offered to come and look after Pampo again and when she arrived was delighted to be welcomed by most endearing squeals.

It was dark when we reached camp, but Elsa appeared within a few moments. She was extremely thin, very hungry and had deep, bleeding gashes and bites on her neck, and also the claw marks of a lion on her back.

While she gnawed at the meat we had brought and I dressed her wounds, she responded by licking me and rubbing her head against mine.

During the night we heard her dragging the carcase down to the river and splashing across with it, and later we heard her returning. Shortly afterwards some baboons gave an alarm and were answered by a lion across the river. Elsa replied from our side with soft moans. Very early in the morning she tried to force her way through the wicker door of the thorn enclosure which surrounds my tent. She pushed her head half-through but then got stuck. Her attempt to free herself caused the door to give way and she finally entered wearing the gate round her neck like a collar. I freed her at once but she seemed restless and in need of reassurance, for she sucked my thumb frantically. Though she was hungry she made no attempt to recover or to guard her 'kill' as she usually does. All she did was to listen intently when any sound came from the direction of the carcase. We were puzzled by this odd behaviour, so George went to investigate what had happened to the 'kill.' He discovered that Elsa had taken it across the river, but the spoor he found on the far side suggested that another lioness had then dragged it about four hundred yards, eaten part of it and afterwards taken the remains towards some nearby rocks. Assuming that this lioness had cubs concealed in the rocks, George did not go on with his search. He observed, however, that beside the spoor of the strange lioness were the pug marks of a lion – and that they were not those of Elsa's husband. The evidence suggested that this lion had not touched the meat but had followed the lioness at some distance, and left the 'kill' to her.

Does this mean that though lions are not of much use to a lioness who is in cub or nursing and therefore handi-capped for hunting, they do make sacrifices for their mate? Had Elsa, though she was hungry, suffering from still unhealed wounds and herself in need of an aunt on account of her pregnancy, gone to the help of a nursing lioness? This was something we could only wonder about.

She was now rather heavy and all exercise had become an effort to her.

Now, when she came with me to the studio she often lay on the table. I was puzzled about this, for though the table is perhaps a cooler place it was certainly a lot harder than my bed, or the soft sand below. During the following days Elsa shared her time between her mate and me. On our last night in camp Elsa made a terrific meal of goat and then, very heavy in the belly, went to join her lion who had been calling for her for many hours. Her absence gave us an excellent opportunity to leave for Isiolo.

When I got home I was horrified at Pampo's appearance. His face had fallen in alarmingly, especially round the eyes and as he dragged himself up to us his bones stuck out. Joan told us that recently his milk consumption had suddenly dropped from two gallons to only six bottles a day. At first she had thought this might be due to teething pains for he was constantly rubbing his gums against anything he could find. He had also pushed his head into his bath tub and sucked up all the water, and on the following day had wished to repeat the performance; but as his digestion was upset Joan had not produced his tub for him. Soon afterwards she had found him trying to satisfy his thirst from a muddy puddle below a waste pipe. His condition had grown worse after this episode and she had called in the vet. He had advised her to feed Pampo on glucose and water only, and had treated him with sulphaguanidine.

After our return Pampo got weaker day by day and in spite of everything we did for him he died, very peacefully, leaning his head against me, just a month after he came to us. I was very upset at losing him as he was a most lovable little creature, but the post-mortem proved that he had pneumonia and diseased intestines as well, so we could not have hoped to save his life.

It was the hottest time of year and there was a severe drought. The tribesmen, who in general avoided the region round Elsa's camp, because it is infested with a type of tsetse fly which is fatal to domestic stock, now offered to pay in order to be allowed to bring their flocks into the reserve. The District Commissioner and George had several meetings with them and did their best to provide a solution to their problem, but, in spite of this, trespassing and poaching increased.

In the second week of November when we were on our way back to Elsa, about ten miles from the camp we noticed a lot of vultures in the trees, and so we went to look for the kill which their presence indicated, and found the body of a baby elephant, scarcely bigger than Pampo.

He had died of spear wounds and had no doubt been killed by Boran tribesmen. In order to win the approval of the girls the young men of this tribe are obliged to engage in 'spear-blooding.' This means that they have to prove their courage by killing an animal belonging to a dangerous species, and unfortunately the fact that the victim itself may be newly born and defenceless in no way invalidates the test.

When we got near Elsa's lie-up we found the spoor of many sheep and goats and the camp site itself patterned with hoof marks. I trembled to think what might have happened to her should she have killed one of the goats which had been grazing so provokingly in what she regarded as her private domain. Later our fears were increased by finding the body of a crocodile close to the river; it had been speared quite recently.

George sent a patrol of Game Scouts to deal with the poachers while he and I went out to look for Elsa.

For some hours we walked through the bush, calling to her and at intervals shooting into the air, but there was no response. After dark a lion began to call from the direction of the Big Rock, but we listened in vain for Elsa's voice.

We had run out of thunder flashes so when it became dark all we could do to let her know that we were there was to turn on the penetrating howl of the air-raid siren, a relic of Mau-Mau days. In the past it had often brought her into camp.

It was answered by the lion; we sounded it again and again he replied, and this strange conversation went on until it was interrupted by Elsa's arrival. She knocked us all over; as her body was wet we realized that she must have swum across the river and had come from the opposite direction to that from which the lion was calling.

She seemed very fit and not hungry. She left at dawn but returned at tea-time when we were setting out for our walk. We climbed up the Big Rock and sat there watching the sun sink like a fire-ball behind the indigo hills.

At first Elsa blended into the warm reddish colour of the rock as if she were part of it, then she was silhouetted against the fading sky in which a full moon was rising. It seemed as though we were all on a giant ship, anchored in a purple-grey sea of bush, out of which a few islands of granite outcrop rose. It was so vast a view, so utterly peaceful and timeless, that I felt as though I were on a 'magic ship' gliding away from reality into a world where man-created values crumble to nothing. Instinctively I stretched my hand towards Elsa who sat close to me; she belonged to this world and only through her were we allowed to glance into a paradise which we had lost. I imagined Elsa in the future playing with her happy little cubs on this rock, cubs whose father was a wild lion; and at this very moment he might be waiting nearby. She rolled on her back and hugged me close to her. Carefully I laid my hand below her ribs to feel whether any life were moving within her, but she pushed it away making me feel as though I had committed an indiscretion. Certainly her nipples were already very large.

Soon we had to return to camp, to the safety of our thorn enclosure, and the lamps and rifles with which we armed ourselves against those dark hours in which Elsa's real life began.

This was the moment at which we parted, each to return to our own world.

When we got back we found that there were a number of Boran poachers in camp who had been rounded up by the Game Scouts. As a Senior Game Warden, one of George's most important tasks is to put down poaching for it threatens the survival of wild life in the reserves.

Elsa kept away during that night and the following day. This worried us as we would rather have had her under our eyes while so many tribesmen and their flocks were around. In the afternoon we went to look for her. As I came near to the rock, I called out to warn her of our approach but got no reply. It was only when we had climbed on to the saddle where we had sat on the previous evening that we suddenly heard an alarming growl, followed by crashes and the sound of wood breaking inside the big cleft below us. We rushed as fast as we could to the top of the nearest rock, then we heard Elsa's voice very close and saw her lion making away swiftly through the bush.

Elsa looked up at us, paused and silently rushed after her mate. Both disappeared in a direction in which we knew there were some Boran with their stock.

We waited until it was nearly dark and then called Elsa. To our surprise she came trotting out of the bush, returned to camp with us and spent the night there, going off only in the early morning.

George went back to Isiolo with the prisoners but left some Game Scouts in camp.

The bush was full of sheep and goats which had straggled away from the flocks and several newly born lambs were bleating piteously. With the help of the Scouts I found them and returned them to their mothers.

The evening was lit by lightning, a sure sign that the rains would start soon. Never had I greeted the first down-pour with such a sense of relief. For this drenching meant that the Boran would return to their pastures and temptation and danger would be removed from Elsa's path.

Fortunately, as she did not like the crowd of Game Scouts who now shared our camp, she spent these last dangerous days on the far side of the river where there were neither Boran nor flocks.

Daily now the parched ground was soaked by showers. The transformation which always results from the onset of the rains is something which cannot be imagined by anyone who has not actually witnessed it.

A few days before we had been surrounded by grey, dry, crackling bush, in which long white thorns provided the only variation in colour. Now, on every side there was lush tropical vegetation decked with myriads of multi-coloured flowers, and the air was heavy with their scent.

As usual the weaver birds made good use of the shoulder-high grass and began to build a colony of nests in two trees which overhung our tents. They seemed to feel safe there. Each morning I woke to the gay chatter of some six hundred weavers all busy building. Most of them were of the yellow black-headed species who make their nests of grass, but I saw also some pairs of red-headed weavers who use twigs. It surprised me that they should have chosen to join the community as they are not usually gregarious.

A pair of the red-heads hung their nest practically above

the entrance to my tent and in spite of our frequent comings and goings tranquilly wove a most beautiful home.

The black-headed weavers began by attaching a grass loop some inches up a twig. Taking this as a starting point, they used their beaks to thread the grass stalks in and out and to make complicated knots. In order to do this they were obliged to hang upside down and they fluttered ceaselessly trying to keep their balance. Busily they flew to and fro collecting suitable fibre but sometimes a bird returning with a long piece of grass dangling from its beak would find that while it had been working a lazy weaver had taken over its nest. They twittered and chirped so much that I sometimes wondered how with all this chattering they found time to use their beaks for weaving, but in fact they completed their nests in two or three days; later hundreds of broken egg shells proved that the young birds had arrived.

The mornings and the evenings were the times at which the parent birds were busiest, though they often continued to chatter long after our lamps were lit. As neighbours they had one inconvenience: although the boys cleaned the canvas of our tents daily they were always coated with droppings.

One morning I found a fledgeling on the ground, chirping unhappily for its mother. I placed it carefully in a nest which had fallen down and then hung this on a twig, vainly hoping that the mother bird would come to her crying chick. From now on such accidents became frequent so I tied a line of nests to our thorn fence in which I placed the orphans. Each nest had one or two occupants, and whenever I came near the chicks opened their triangular, yellow-lined beaks and demanded food.

Luckily we were invaded by sugar ants who hide their juicy larvæ in dark places, and these provided an ample supply of food. With the help of forceps I dropped the grubs down the chicks' throats. I had to be very quick about repeating the dose for the little birds nearly fell out of their nests in their attempts to get one more grub, and several of the older fledgelings who could already make a little use of their wings parachuted on to the ground. Others I had to hold in my hand and reassure for a long time before they would accept any food. Most of them were eventually able to look after themselves but a few I could not save, and I was miserable when I witnessed the long struggle which ended in their death.

In camp evening is the time that I like best, for it is then that one becomes aware of the monotonous vibrations of the crickets and the rumble of the elephants, the hum of the bush, pierced occasionally by the cry of some nocturnal animal.

It is then too that one sees the great belt of light, some ten feet wide, formed by thousands upon thousands of fireflies whose green phosphorescence bridges the shoulder-high grass. The fluorescent band composed of these tiny organisms lights up and goes out with a precision which is perfectly synchronized, and one is left wondering what means of communication they possess which enables them to co-ordinate their shining as though controlled by a mechanical device.

I had spent many rainy seasons in camp but never before had I seen such a brilliant display.

When George returned he brought a zebra for Elsa. This was a special treat. As soon as she heard the vibrations of the car she appeared, spotted the 'kill' and tried to pull the carcass out of the Landrover. Then, finding it too heavy for her, she walked over to where the boys were standing and jerking her head at the zebra made it plain that she needed help. They hauled the heavy animal a short distance amid much laughter and then waited for Elsa to start her meal. To our astonishment, although zebra was her favourite meat she did not eat but stood by the river roaring in her loudest voice.

We presumed that she was inviting her mate to join in the feast. This would have been good lion manners, for according to the recorded habit of prides, whilst the females do most of the killing, they then have to wait to satisfy their hunger until the lion has had his fill.

The next morning, 22nd November, she swam across the heavily flooded river, came up to the zebra and roared repeatedly in the direction of the rocky range which is on our side of the river.

I noticed that she had a deep gash across one of her front paws, but she refused to have it dressed, and after she had eaten as much as she could, she went off towards the rocks.

That night it rained for eight hours, and the river turned into a torrent which it would have been very dangerous for Elsa to cross even though she is a powerful swimmer. I was therefore very pleased to see her in the morning returning from the Big Rock.

Her knee was very swollen and she allowed me to attend to her cut paw.

I noticed that she had great difficulty in producing her excrement and when I inspected the fæces I was surprised to see a rolled-up piece of zebra skin which when unfolded was as large as a soup plate. The hair had been digested but the hide was half an inch thick. I marvelled at the capacity of wild animals to rid themselves of such objects without suffering any internal injury.

For several days she divided her time between us and her lion.

When George returned from a patrol he brought Elsa a goat. Usually she dragged her 'kill' into his tent, presumably to avoid the trouble of having to guard it, but this time she left it lying beside the car in a spot which could not be seen from the tent. During the night her mate came and had a good feed; we wondered whether this was what she had intended.

Next evening we took the precaution of placing some meat at a certain distance from the camp, for we did not want to encourage him to come too close.

Soon after dark we heard him dragging it away and in the morning Elsa joined him.

We were now faced with a problem. We wanted to help Elsa, who was increasingly handicapped by her pregnancy, by providing her with regular food, but we did not wish to interfere with her relations with her mate by our continued presence in the camp. He had a good right to resent this, but did he in fact object to us? On the whole, we thought that he did not, and I think we were justified in our opinion for, during the next six months, though we did not see him, we often heard his characteristic ten or twelve whuffing grunts and recognized his spoor, which proved that he remained Elsa's constant companion.

Though he still kept out of our sight, he had become bolder and bolder, but an extraordinary kind of truce seemed to have been established between us. He had come to know our routine as intimately as we had come to know his habits. He shared Elsa's company with us and we thought that in return he could fairly expect an occasional meal as compensation.

In view of his attitude we stilled our qualms of conscience and stayed on.

One afternoon walking with Elsa through the bush we came upon a large boulder with a crack in it. She sniffed cautiously, pulled a grimace and did not seem anxious to go closer to it. Next we heard a hissing and, expecting a snake to appear, George held our shotgun ready; but what emerged from the crack was the broad head of a monitor lizard who soon wriggled out into the open. He was an enormous size, about five feet long and nearly a foot broad and he had blown himself up to his fullest capacity. He extended his neck, moved his long forked tongue rapidly and lashed out with his tail so violently that Elsa thought it wise to retreat.

Sitting at a safe distance, I admired his courage; although he had no means of defence except his threatening appearance and thrashing tail, which he used like a crocodile, he chose to come out and face the danger, rather than find himself trapped in the crack.

On our way home we climbed to the top of Elsa's favourite rock and took some photographs of her. She posed beautifully until she heard her lion calling from just below, then she went down the rock into a steep ravine. Watching her, I was amazed that such a heavy animal should be able to keep its balance on the almost vertical rock face.

For a few days we saw little of her but as we often heard her lion roaring and frequently saw his pug marks we did not worry.

On the 1st December in the afternoon she came back and accompanied us when we walked to a rain pool; there she lay at the water's edge while I sat next to her and killed the tsetse flies which, in the failing light, were beginning to bite. While doing so, I read the 'bush newspaper' in terms of the freshly imprinted spoor which surrounded the pool.

Suddenly I heard George give a whistle and looking up saw a herd of some twenty buffalo cows, many of them followed by calves, making their way to the water.

Elsa stared at the herd, raised herself very cautiously to a crouching position, with her head on her paws, and then suddenly rushed at top speed towards the herd. There was a thundering noise and the crash of breaking wood as the buffaloes bolted with Elsa in hot pursuit.

We ran after her as fast as we could and found her facing a thicket, panting hard. From within the bush came the angry snorting of the buffaloes; they had evidently rallied and were preparing to defend their young. A moment later several enraged cows charged Elsa who, recognizing her limitations, withdrew, keeping in line with George, myself and Makedde. Then she made a series of quick thrusts forward, but returned equally fast to her support.

George waited until the herd was within about fifteen yards of us, then he and Makedde shouted and each waved one arm, holding his rifle in the other. The animals were puzzled by this strange performance and after a moment of indecision turned and made off.

After a while we followed, but we took good care to make certain that no buffalo was waiting to ambush us for they are notoriously dangerous creatures.

Next morning George had to leave; I stayed on and Elsa

spent three days in camp with me in spite of the continual calling of her mate.

One evening she looked towards the river, stiffened and then rushed into the bush. A tremendous barking of baboons ensued, till it was silenced by her roars. Soon she was answered by her lion – he must have been only about fifty yards away. His voice seemed to shake the earth and increased in strength. From the other side Elsa roared back. Sitting between them, I became a little anxious in case the loving pair should decide to come into my tent, for I had no meal to offer them. However, in time they appeared to have roared themselves hoarse. Their whuffings died away and no further sound came from the bush except for the buzzing of insects. Luckily on the following evening George returned with a goat for Elsa.

During the rainy season the atmosphere is so full of humidity that raw meat goes bad quickly, often in less than two days. In order to preserve the remains of Elsa's meals for as long as possible we improvised a bush fridge! We wrapped a lot of foliage around the meat to prevent flies from laying their eggs in it and then we hung it from the branch of a shady tree about two feet above the ground.

This tree stood at a short distance from the home of a monitor, which was occupied at the time by an adult and a youngster who had just shed his skin. One morning, on my way to the studio, I caught sight of the old monitor looking greedily at the bundle which was dangling just outside his reach. He saw me and beat a hasty retreat, but soon afterwards I heard a rustling in the leaves and there was the young monitor. I kept absolutely still; making several detours, he came to within a few feet of me. When he seemed to have satisfied himself that I was harmless he went home. Next the old monitor, who had obviously sent out his offspring as a scout, reappeared and advanced stealthily towards the meat. For a time he sat beneath the bundle contemplating it, then he jumped at it, touched it and fell back. After this he repeated his jumps until he finally got a grip and disappeared into the larder.

I gave him time to have a good meal, then I clapped my hands, and immediately he fell to the ground with a plop. He looked quite ridiculous, for his mouth was still full of meat which dangled out on both sides. Instead of bolting he sat motionless staring at me as though he hoped to mesmerize me. As I did not move he was evidently reassured, for though he never took his eyes off me, he began to gulp as fast as he could and did not waddle off until he had finished his mouthful.

2
The Birth of the Cubs

It was now nearly mid-December and we believed that the cubs might arrive at any moment.

Elsa was so heavy that every movement seemed to require an effort; if she had been living a normal life she would certainly have taken exercise, so I did my best to make her go for walks with me, but she kept close to the tents. We wondered what place she would choose for her delivery and even thought that since she had always considered our tent as her safest 'den' the cubs might be born in it.

We therefore prepared a feeding bottle and laid in some tinned milk and some glucose, and I read all the books and pamphlets I could find on animal births and possible complications.

Since I had no experience of midwifery I felt very nervous and also asked advice of a veterinary surgeon. In order to judge how far Elsa was advanced in her pregnancy I pressed my hand gently against her abdomen just below her ribs. I could not feel any movement and wondered whether we had been mistaken about the date at which she had been mated.

The river was now in flood and George and I decided to walk three miles downstream to look at some cataracts which are very impressive when the water is high. Elsa watched our departure from the top of the Landrover. She made no attempt to join us and looked sleepy. The bush we had to go through was very thick and as we walked I wished she were with us to warn us of the approach of buffalo and elephant, for droppings proved that they must be close by.

The cataracts were a magnificent sight, the foaming water cascading through the gorges, thundering across the rocks and then flooding out into deep whirlpools.

On our way back, as soon as I was out of earshot of the cataracts, I heard Elsa's familiar *hnk-hnk* and soon saw her trotting along the path as quickly as she could to join us. She was covered with tsetse flies, but she greeted us most affectionately before she flung herself on the ground, and tried to rid herself of the flies by rolling.

I was very touched that she had made the effort to join us, the more so that though her lion had roared desperately for her during the whole of the previous night and had gone on doing so until nine in the morning, she had made no attempt to join him.

This was very gratifying but it also reminded us of our fear that her lion might get tired of sharing her with us. It had taken us a very long time to find a mate for her; it would be unforgivable if our interference now caused him to leave her. We wanted her cubs to grow up as wild lions and to do this they needed a father.

We decided to go away for three days. It was of course a risk, for the cubs might be born during this time and Elsa might need us, but we thought the danger that her lion might desert her the greater of the two evils – so we left.

We returned on the 16th December and found a very

hungry Elsa waiting for us. For two days she remained in camp; possibly frequent thunderstorms made her reluctant to leave its shelter. She did, however, to our surprise, take a few short walks, always to the Big Rock, but returned quickly. She ate unbelievably and we felt that she was stocking up a reserve for the days that lay ahead.

On the night of the 18th December she crept in the dark through the thorn fence which surrounded my tent and spent the night close to my bed. This was something which she had very rarely done, and I took it as a sign that she felt that her time was near.

The next day when George and I went for a walk Elsa followed us, but she had to sit down at intervals panting and was plainly in great discomfort. When we saw this we turned back and walked very slowly. Suddenly to our astonishment she turned off into the bush in the direction of the Big Rock.

She did not return during that night, but in the morning we heard her calling in a very weak voice. We thought this meant that she had had her cubs and went out to trace her spoor. This led us close to the rock but the grass was so high that we lost track of her. The rock range is about a mile long and though we searched for a long time we could not discover where she was.

We set out again in the afternoon and eventually we spotted her through our field-glasses. She was standing on the Big Rock and from her silhouette we saw that she was still pregnant.

We climbed up and found her lying close to a large boulder which stood at the top of a wide cleft in the rock; near to it there was some grass and a small tree provided shade. This place had always been one of Elsa's favourite 'lookouts' and we felt that it would make an ideal nursery, since inside the cleft was a rainproof and well-protected cave.

We left her to take the initiative and presently she came slowly towards us, walking very carefully and obviously in pain. She greeted us very affectionately, but I noticed that blood was trickling from her vagina, a sure sign that her labour had started.

In spite of this she went over to Makedde and the Toto, who had remained behind, and rubbed her head against their legs before she sat down.

When I came near her she got up and moved to the edge of the rock, and remained there with her head turned away from us. It seemed to me that she chose this precipitous position to make sure that no one could follow her. At intervals she came back and rubbed her head very gently against mine and then walked determinedly back to the boulder making it plain that she wished to be left alone.

We went a short distance away and for half an hour watched her through our field-glasses. She rolled from side to side, licked her vagina and moaned repeatedly. Suddenly she rose, went very carefully down the steep rock face and disappeared into the thick bush at its base.

Since there was nothing we could do to help her, we went back to camp. After dark we heard her lion calling; there was no reply.

I lay awake most of the night thinking about her and when, towards morning, it started to rain my anxiety increased and I could hardly bear to wait till it was light to go out and try to discover what had happened.

Very early, George and I set out; first we followed the spoor of Elsa's lion. He had been close to the camp, had dragged off the very smelly carcase of the goat which Elsa had not touched for three days, and had eaten it in the bush. Then he had walked to the rock near to the place where we had seen Elsa disappear.

We wondered what we should do next. We did not want our curiosity to bring any risk to the cubs and we were aware that captive lionesses who have been disturbed soon after giving birth to cubs have been known to kill their young. We also thought that her lion might be very near, so we decided to stop our search; instead George went off and shot a large water buck to provide Elsa and her mate with plenty of food.

I, in the meantime, climbed the Big Rock and waited for an hour, listening for any sound which might give us a clue to Elsa's whereabouts. I strained my ears but all was still; finally I could bear the suspense no longer and called. There was no answer. Was Elsa dead?

Hoping that the lion's spoor might lead us to her we took up his tracks where we had left them and traced them till they reached a dry watercourse near the rock. There we left his meal thinking that if he came for it this might help us to find Elsa.

During the night we heard him roaring in the distance and were therefore surprised next morning to find his pug marks close to the camp. He had not taken any of the meat that was close to the camp but had gone to the 'kill' we had left for him near the rock. This he had dragged for at least half a mile through most difficult terrain, across ravines, rocky outcrops and dense bush. We had no wish to disturb him at his meal, so we set about looking for Elsa, but found no trace of her. After returning to the camp for breakfast we went out again and suddenly, through our field-glasses, saw a great flock of vultures perched on the trees which grew around the spot where we thought that the lion had made his meal.

Assuming that he had finished by now, we approached the place and as we came near to it found every bush and tree loaded with birds of prey. Each was staring at the dry watercourse and there was the carcase lying out in the hot sun. Since the meat was in the open and yet the vultures did not leave their perches we concluded that the lion was guarding his 'kill'. As far as we could see he had not touched it, so we thought that Elsa too might be close by and that her gallant mate had dragged the four-hundred-pound burden this long distance for her benefit. We felt it would be unwise to continue our search and went back to camp for lunch after which we set out again.

When we saw that the vultures were still on the trees, we circled the place down-wind and approached it very cautiously from the high ground.

George, Makedde and I had just passed a very thick bush which overhung a deep crack in the ground when I suddenly had a strange uncomfortable feeling. I stopped and looking back, saw the Toto, who was close behind me, staring intently at the bush. Next there was a terrifying growl and the sound of snapping branches; a second later all was quiet again – the lion had gone. We had passed within six feet of him. I think that my sense of uneasiness must have been due to the fact that he had been watching our movements with great intensity. When the Toto stooped to see what was in the bush he couldn't stand it and went off. They had actually looked straight into each other's eyes and the Toto had seen his big body disappearing into the deep crack. Feeling we had been very lucky, we went home and left three lots of meat in different places before night fell.

As soon as it was light we went to inspect the deposits; all of them had been taken by hyenas.

By the river we found the spoor of Elsa's mate, but there was no sign of her pug marks. All the little rain pools had dried up long ago and the river was the only place where she could quench her thirst; the absence of any trace of her was very worrying. Eventually we found, close to the spot where three days before we had last seen her, a few pug marks which could have been hers, though this was not certain. Full of hope, we made a thorough search along the base of the Big Rock, but in vain.

Since the vultures had now gone we were left with no clue to her whereabouts.

Again we put out meat close to the rock and near to the camp. In the morning we found that Elsa's lion had dragged some of it to the studio and eaten it there, while the rest had been disposed of by hyenas.

It was now four days since we had seen Elsa and six since she had eaten anything, unless she had shared the water buck with her mate.

We believed that she had given birth to the cubs on the night of the 20th December and we did not think that it could be a coincidence that her lion, who had not been about for days, had reappeared on that night and remained close to the rock ever since; which was most unusual.

On Christmas Eve George went to get a goat while I continued the fruitless search and called to Elsa without getting any answer.

It was with a heavy heart that I prepared our little Christmas tree. In the past I had always improvised one;

sometimes I took a small candelabra euphorbia, from whose symmetrical branches I hung tinsel chains and into whose fleshy fibre I stuck candles; sometimes I used an aloe with its wide-spreading sprays of flowers, sometimes a seedling of the thorny balanitis tree, which is very ornamental and has splendid spikes on which to hang decorations. When I could find nothing else I filled a dish with sand, stuck candles into it and decorated it with whatever plants I could pick in our semi-desert surroundings.

But tonight I had a real little tree complete with glittering tinsel branches, sparkling decorations and candles. I placed it on a table outside the tents which I had covered with flowers and greenery. Then I collected the presents which I had brought for George, Makedde, Nuru, Ibrahim, the Toto and the cook and the sealed envelopes containing money for the boys on which I had painted a Christmas tree branch. There were also packets of cigarettes and dates and tins of milk for them.

I changed quickly into a frock and by then it was dark enough to light the candles. I called the men, who came dressed up for the occasion, grinning but a little shy, for never before had they seen a Christmas tree of this kind.

I must admit to having been myself deeply moved when I saw the little silver tree sparkling in the vast darkness of the surrounding bush, bringing the message of the birth of Christ.

On Christmas Eve I always feel like a small child. To break the tension, I told the men about the European custom of celebrating Christmas Eve with a tree. After I had given them their presents, we all gave three cheers for 'Elsa – Elsa, Elsa.' The sound seemed to hang on the air and I felt a lump rise in my throat – was she alive? Quickly

I told the cook to bring in the plum pudding which we had brought from Isiolo and then to pour brandy over it and light it. But no bluish flame arose, for our Christmas pudding was a soggy mass which had a distinct smell of Worcester sauce. Certainly the cook had never before been in charge of such a ritual; he had paid no attention to my instructions and had remained fixed in his belief that George so loved his Lea and Perrins, that it must be appropriate to souse even the plum pudding with it.

We were not, however, the only ones to be disappointed in our Christmas dinner. We had hung a goat carcase out of the reach of predators, which we would lower if Elsa appeared. After we had gone to bed we heard her lion grunting and growling by the tree and performing all sorts of acrobatics. He went on for a long time and then retired exhausted.

Early on Christmas morning we went in search of Elsa. We followed the lion's spoor across the river, and again screened the bush all round the spot to which he had dragged the water buck. After hours of fruitless tracking we came back for breakfast. During the morning George shot at an aggressive cobra which we found close to the camp.

Later we set out once more for the rocky range; something seemed to tell us that if Elsa were still alive that was were she was. We wriggled through dense bush and I crept hopefully into every crevice trying to prevent myself from expecting to find Elsa dead but hidden from the vultures by the impenetrable thorn thickets.

When we were all tired out we sat down to rest in the shade of an overhanging rock and discussed every possible fate which might have overtaken Elsa. We were very depressed and even Nuru and Makedde spoke in subdued voices.

We tried to cheer ourselves up by quoting cases of bitches who would not leave their litter for the first five or six days because they had to keep them warm, feed them and massage their bellies to help their digestive functions to start working. Indeed, we had expected Elsa to have a rather similar reaction, but this did not account for the absence of any trace of her. Also, bitches do occasionally go and visit their masters even during this first period after their delivery and as Elsa had shown more attachment to us than to her mate up to the time at which her labour began, it seemed improbable and ominous that the fact of giving birth to cubs should have caused her to go completely wild.

At midday we returned to camp and began a very gloomy and silent Christmas meal.

Suddenly there was a swift movement and before I could take in what was happening Elsa was between us sweeping everything off the table, knocking us to the ground, sitting on us and overwhelming us with joy and affection.

While this was going on the boys appeared and Elsa gave them too a full share of her greetings.

Her figure was normal again, she looked superbly fit but her teats were very small and apparently dry; round each was a dark-red circle some two inches wide. Cautiously I squeezed a teat; it produced no milk. We gave her some meat which she immediately ate. Meanwhile, we discussed many questions. Why had she come to visit us during the hottest part of the day, a time when normally she would never move? Could it be that she had chosen it deliberately because it was the safest time to leave the cubs since few predators would be on the prowl in such heat; or, had she heard the shot which George had fired at the cobra and had she taken it as a signal to her? Why were her teats small and dry? Had she just suckled the cubs? But this would not seem to explain why her milk glands which had been so big during her pregnancy had now shrunk to their normal size. Had the cubs died? And whatever had happened, why had she waited five days before coming to us for food?

After she had had a good meal and drunk some water she rubbed her head affectionately against us, walked about thirty yards down the river, lay down and had a doze. We left her alone, so that she should feel at ease. When I looked for her at tea-time she had gone.

We followed her spoor for a short way; it led towards the rock range, but we soon lost it and returned none the wiser about her cubs. However, now that we were reassured about Elsa our morale was restored.

During the night we heard her lion calling from the other side of the river, but she did not answer him.

Next day we began to worry about the cubs. If they were alive was their mother able to suckle them from those dry teats? We tried to comfort ourselves by saying that the red rings round them were probably due to blood-vessels being broken by suckling, but we were very anxious because we had been warned by zoo authorities that hand-reared lionesses often produce abnormal cubs which do not live, and indeed one of Elsa's sisters had suffered such a misfortune. We felt we just must know about the cubs and rescue them if necessary. So the next morning we searched for five hours, but we did not find so much as a dropping or a crushed leaf, let alone any spoor to show where Elsa's nursery was.

We carried on equally unsuccessfully in the afternoon. While plodding through the bush George nearly stepped on an exceptionally large puff adder and was lucky to be able to shoot it just before it could strike.

Half an hour later we heard Ibrahim popping off a gun, a signal that Elsa had arrived in camp.

Obviously she had responded to the shot with which George had dispatched the puff adder.

She was most affectionate to us when we got back, but we were alarmed to observe that her teats were still small and dry. Ibrahim, however, assured us that when she had arrived they and her milk glands had been enormous, hanging low and swinging from side to side.

He also told us that her behaviour had been very unusual. When he fetched the gun from the kitchen which was in the direction from which she had come she dashed angrily at him. Possibly she thought he was going to her cubs. Later when he went to the studio to collect her meat which was hanging there in the shade, she had prevented him from touching her 'kill.' After this she had settled on the Landrover and it was then that Ibrahim noticed that her teats and glands had shrunk to their normal size. She had, he said, 'tucked them up,' and he told us that camels and cattle can withhold their milk by retracting their teats. If then their owner insists on getting milk he is obliged to tie the animal to a tree and apply several tourniquets; these have the effect of raising the pressure of the blood in the muscles until it reaches a point when they automatically relax and it becomes possible to start milking. We wondered whether such a retraction explained the peculiar state of Elsa's teats. Was it not possible that a lioness might be capable of a similar reaction and would contract her teats when hunting? Certainly if she could not do this she would be greatly handicapped by her heavy undercarriage, and besides this her teats might be injured by the thorny bush.

While we were asking ourselves these questions, Elsa, having eaten enormously, had settled down and showed no intention of returning to her cubs.

This alarmed me because it was getting dark and the worst moment to leave them alone.

We tried to induce her to return to them by walking along the path down which she had come. She followed us reluctantly, listening alertly in the direction of the rock, but soon returned to camp. We wondered whether she might be afraid that we would follow her and find her cubs. Meanwhile she went back to her meal and it was only after she had methodically cleaned up every scrap of it that, much to our relief, she disappeared into the dark. Very likely she had waited till there was no light to make sure we could not follow her.

We were now convinced that she was looking after her cubs. But after the warnings we had had from the zoo experts we could not be happy until we had seen for ourselves that they were normal.

We made one more unsuccessful search before our return to Isiolo where we spent the last three days of December. On our way back to camp we nearly collided with two rhino and then met a small herd of elephant. We had no choice but to rush past them, hoping we should 'make it,' but the big bull of the herd took umbrage and chased us for quite a long way. I did not enjoy this as elephants are the only wild animals which really frighten me.

We hooted several time before we reached camp to let Elsa know we were arriving and found her waiting for us on top of a large boulder at the point at which the track passes the end of the Big Rock.

She hopped in among the boys at the back of the Land-rover, then she went to the trailer in which there was a dead goat. I had rarely seen her so hungry.

I noticed at once that her teats were still small and dry; I squeezed them, but no milk came. We thought this a bad sign and after she had spent seven hours in camp, eating and hopping on and off the Landrover, we began to be afraid that she no longer had any cubs to look after. She only left us at two in the morning.

Very early we set out and followed her spoor which led towards the Big Rock. Close to it was what seemed to us an ideal home for a lioness and her family. Very large boulders gave complete shelter and they were surrounded by bush that was almost impenetrable. We made straight for the topmost boulder and from it tried to look down into the centre of the 'den.' We saw no pug marks but there were signs that some animal had used it as a lie-up.

Nearby we observed some old blood spoor. This was very close to the place where we had seen Elsa in labour, so we thought that she had perhaps given birth to the cubs there. On the other hand, we had been within three feet of it on one of our previous searches and it seemed almost impossible that Elsa should have been there hiding her cubs and not made us aware of her presence.

As though to prove that we were wrong in thinking this, after we had called loudly for half an hour, she suddenly appeared out of a cluster of bush only twenty yards away. She seemed rather shocked at seeing us, stared and kept silent and very still as though hoping we would not come nearer.

Perhaps we were so close to her nursery that she thought it better to appear and so prevent us from finding it. After a few moments, she walked up to us and was very affectionate to George, myself, Makedde and the Toto, but never uttered a sound. To my relief I saw that her teats were twice their normal length and that the hair around them was still wet from suckling.

Soon she went slowly back towards the bush and stood, for about five minutes, with her back turned towards us listening intently for any sound from the thicket. Then she sat down, still with her back turned to us. It was as though she wanted to say to us: 'Here my private world begins and you must not trespass.'

It was a dignified demonstration and no words could have conveyed her wishes more clearly.

We sneaked away as quietly as we could, making a detour in order to climb to the top of the Big Rock. From it we looked down and saw her sitting just as we had left her.

Obviously she had got our scent, knew just what we were doing and did not intend to let us discover her lie-up.

This made me realize how unaware we had been, in spite of our intimacy with Elsa, of the reactions of wild animals. It amused me to remember how we had prepared ourselves against the possibility of the cubs being born in our tent and how we had flattered ourselves that Elsa regarded it as the place in which she felt safest. Although the spoor we had recently found had all led towards the lower rock, we thought it possible that the cubs had been born in the boulder hideout and that later Elsa had moved them about thirty yards to where they now were.

If this were the case she had probably made the move after the rains stopped – for while the boulder lie-up was rainproof, the new one was not, though otherwise it was an ideal nursery.

We decided that we must respect Elsa's wishes and not try to see the cubs until she brought them to us, which we felt sure she would do one day. I determined to stay on in camp in order to provide her with food so that she would have no need to leave her family unguarded for long periods while she went out hunting for them. We also decided to take her meals to her, so as to reduce the time during which she had to desert the cubs.

We put our plan into immediate operation and that afternoon went by car close to her lie-up. We knew that Elsa would associate the vibrations of the engine with us and with food.

As we neared the place where we had last seen her we started to call out – 'Maji, Chakula, Nyama' – Swahili words, meaning water, food, meat, with which Elsa was familiar.

Soon she came, was as affectionate as usual and ate a lot. While she had her head in a basin, which we had sunk in the ground to keep it steady and was busy drinking, we went off. She looked round when she heard the engine start but made no move to follow us.

Next morning we took her her day's ration but she failed to turn up, nor was she there when we went again in the afternoon. During the night a strange lion came to within fifteen yards of our tent and removed the remains.

After breakfast we followed his spoor which led to the Big Rock and pug marks there showed that another lion had been with him. We hoped that Elsa was enjoying their company and that perhaps they were helping her with her housekeeping.

We went down to the river to see whether she had left any spoor there. She had not, but soon afterwards George, who was going to fetch another goat, met her near her rock. She was very thirsty, the aluminium drinking basin had gone and we wondered whether the other lions had stolen it. On his return George fed her and from her appetite he thought it unlikely that the lions had provided her with any of the food they had stolen.

Later in the day George went off to Isiolo. Elsa stayed in camp with me till the late afternoon, then I saw her sneak into the bush upstream and followed her. Obviously she did not wish to be observed, for when she caught my scent, she pretended to sharpen her claws on a tree. Then as soon as I turned my back on her, she jumped at me and knocked me over, as though to say, 'That's for spying on me!' Now it was my turn to pretend that I had only come to bring more meat to her. She accepted my excuse, followed me and began eating again. After this nothing would induce her to return to the cubs until long after night had fallen and I was reading in my tent and she felt certain that I would not be likely to follow her.

During the following days I went on taking food to the spot near to which we believed the cubs to be. Whenever I met Elsa on these occasions, she took great pains to conceal the whereabouts of her lie-up, often doubling back on her tracks, no doubt to puzzle me.

One afternoon when I was passing the Big Rock I saw a very strange animal standing on it. In the dim light it looked like a cross between a hyena and a small lion. When it saw me it sneaked off with the gait of a cat. It had obviously spotted the cubs and I was much alarmed. Later when I brought up some food, Elsa came at once when I called her; she seemed unusually alert and was rather fierce to the Toto. I left her still eating on the roof of my truck. It was there that we placed the meat in the evening to keep it out of the reach of predators, few of which would be likely to risk jumping on to this unknown object, even if they were capable of doing so. I did not know what to do for the best. If I continued to leave food close to Elsa's nursery, would it not attract predators? Alternatively, if I kept the meat in camp and Elsa had to desert her cubs to come and fetch it, might they not be killed while she was absent? Faced with these two unsatisfactory choices, I decided, on balance, to go on providing food near to her lie-up. When I did so on the following evening, I heard the growls of several lions close to me and Elsa appeared to be both very nervous and very thirsty.

After this I made up my mind that in spite of her disapproval I had better find out how many cubs there were and whether they were all right. I might then be able to help in an emergency. On the 11th of January I did an unpardonable thing. I left a Game Scout (Makedde was ill) with the rifle on the road below and accompanied by the Toto, whom Elsa knew well, I climbed the rock-face calling repeatedly to warn her of our approach. She did not answer. I told the Toto to take off his sandals so as not to make any noise.

When we had reached the top we stood on the edge of the cliff and raked the bush below with our field-glasses. Immediately under us was the place from which Elsa had emerged that first time, when we had surprised her and she had stood on guard.

Now, there was no sign of her, but the place looked like a well-used nursery and was ideal for the purpose.

Although I was concentrating very hard on my examination of the bush below us I suddenly had a strange feeling, dropped my field-glasses, turned and saw Elsa creeping up behind the Toto. I had just time to shout a warning to him before she knocked him down. She had crept up the rock behind us quite silently and the Toto only missed toppling over the cliff by a hair's breadth and that mainly because his feet were bare which gave him the chance of getting a grip on the rock.

Next Elsa walked over to me and knocked me over in a friendly way, but it was very obvious that she was expressing annoyance at finding us so close to her cubs.

After this demonstration, she walked slowly along the crest of the rock, from time to time looking back over her shoulder to make sure that we were following her. Silently she led us to the far end of the ridge. There we climbed down into the bush. As soon as we were on level ground she rushed ahead, repeatedly turning her head back to confirm that we were coming.

In this way, she took us back to the road, but she made a wide detour, presumably to avoid passing near the cubs. I interpreted her complete silence to a wish not to alarm them or to prevent them from emerging and following us.

When we walk together I usually pat Elsa occasionally and she likes it, but today she would not allow me to touch her and made it clear that I was in disgrace. Even when she was eating her dinner on the roof of the car back in camp, whenever I came near her she turned away from me.

She did not go to the cubs until it was dark.

Now George came up from Isiolo and we changed guard. Elsa had made me feel that I could do no more spying on her; George had not had the same experience, so he had fewer inhibitions. My curiosity was immense and I felt that it would be a happy compromise if he did 'the wrong thing' and I were to profit by his misdeed.

3

We See the Cubs

One afternoon, while I was at our home in Isiolo a hundred miles away, George crept very quietly up Elsa's Big Rock and peered over the top.

Below he saw her suckling two cubs and as her head was hidden by an overhanging rock, he felt sure that she had not seen him. Having seen the family, George went back to camp and collected a carcase.

We had brought a number of goats into camp so as to supply Elsa with food and thus prevent her from having to desert the cubs while she went hunting for them, and in doing so risk their being killed by predators.

After depositing the food nearby, George waited to see what would happen. Elsa did not come to fetch the meat. This made him feel guilty. The meat we had put near to where we imagined her to be had always been eaten. Did the fact that on this day she refused to go near the 'kill' indicate that she was aware that George had spied on her? When, during the following day, she failed to come to camp, George feared that this might be the case. However, at nightfall she arrived and was so ravenously hungry that she even condescended to eat a Dik Dik, which she usually despises. It was all he had been able to find for her, and I did not return from Isiolo till a few days later, having picked up a new supply of goats *en route*.

How thrilled I was upon arrival to hear the good news!

George left for Isiolo the next day and I took on the task of supplying Elsa with the vast quantity of food she needed while suckling the cubs.

I noticed very soon that while she was as affectionate as ever to me, even allowing me to hold bones while she gnawed at them, and equally affectionate to George when he was there, she had become much more reserved in her attitude towards Africans, and even her old friends Nuru and Makedde who had known her since she was a cub were not allowed to be as familiar with her as they had been before the arrival of her family.

One day Elsa caused me a lot of anxiety by arriving in camp soon after lunch and showing no sign of returning to her family after she had had her meal. When it got dark I tried to induce her to go back to them by walking in their direction accompanied by the Toto.

She began by following us but after some time turned into the bush, went forward a hundred yards and then sat down with her back towards us blocking our way.

Nothing would budge her, so we took the hint and retired hoping that once we were out of sight she would rejoin her cubs.

On the following day she again showed that she was determined to conceal the whereabouts of the cubs. The Toto and I were taking an afternoon stroll past the Big Rock, walking very quietly. Suddenly Elsa appeared, rubbed her head against my knees and then led us silently away from the Big Rock, where the cubs were, towards a collection of small rocks which we call the Zom rocks.

She crept in and out of crevices, passed between narrow clefts, and seemed to enjoy making us struggle through the most awkward places. If we fell behind she waited for us, often jerking her head as though to show that she expected us to follow her. Finally I sat down, partly to show that I knew I was being fooled.

After this Elsa left the Zom rocks and led us through thorny thickets and boulders, farther and farther away from her lie-up.

At times she sniffed long and portentously at promising places and seemed to be teasing us by trying to make us think that she was taking us to the cubs. Later we passed a place where she was in the habit of ambushing me. I was tired and not prepared to be knocked down, so I made a detour. When she realized this, she emerged from her hideout looking very dignified but obviously disappointed at being cheated of her fun.

The brief sight George had had of the two suckling cubs had not given him time to discover whether they were normal or not and of course he could not tell whether there might be others hidden from his view. So on the afternoon of the 14th January, when Elsa was in camp feeding, he crept off to the Zom rocks, while I kept her company.

For two days she had been constantly in this area, so we supposed that she had changed the place of the nursery.

George climbed up to the top of the centre rock and inside a cleft saw three cubs; two were asleep, but the third was chewing at some Sansevieria; it looked up at him but as its eyes were still blurred and bluish he did not think that it could focus well enough to see him.

He took four photographs but did not expect to get good prints for the cleft in which the cubs lay was rather dark. While he was doing this the two cubs who had been sleeping woke up and crawled about. It seemed to him that they were perfectly healthy.

When he came back to camp and told me the excellent news Elsa was still there and quite unsuspicious.

At dusk we drove her near to the Zom rocks. But only after we had tactfully walked away and she was reassured by hearing our voices fading into the distance did she jump off the Landrover and, presumably, rejoin the cubs.

George now went back to Isiolo. The morning after he had left I herd Elsa's mate calling from the other side of the river but I listened in vain for her reply. In the afternoon, however, she roared very loudly quite near to the camp and went on doing so until I joined her. She seemed overjoyed at seeing me and came back to camp with me, but ate very little and went off when it became dark.

During the next two days she did not turn up, but her mate called to her repeatedly during both nights. On the third day, while I was having breakfast, I heard a terrific roaring coming from the direction of the river. I rushed down to it and saw Elsa standing in the water making as much noise as she could.

She looked very exhausted and soon turned back and disappeared into the bush on the opposite bank. I was puzzled by her odd behaviour. At tea-time she came into camp for a hurried meal and then disappeared. On the following day she did not come, but that night I was woken up by the sound of a large animal thumping at my truck. It stood just outside my thorn enclosure. At night we used it as a goats' stable to protect them from predators. Evidently a lion was trying to get at the goats. I did not think it could be Elsa for she usually gave a characteristic low moan, so I suspected her mate.

I listened intently but, believing that a wild lion was close, I did not make a sound. However, when the banging and rattling increased to such a pitch that I feared the car might be destroyed I flashed a torch. The only result was still heavier thumping.

Suddenly I heard Elsa's mate calling from across the river; this proved it must be she who was attacking the truck. She was plainly furious, but it was dark and I did not want to call the boys to let me out of my enclosure, particularly as I feared that her battering might induce her mate to come to help her. All I could do was to shout, 'ELSA, NO – NO.' I had little hope of being obeyed and was very surprised when she at once stopped her attack and soon left the camp.

On the following afternoon – it was 2nd February – while I was writing in the studio (a place on the river-bank overhung by the branches of a large tree where I work), the Toto came running to tell me that Elsa was calling in a very strange voice from the the other side of the river. I went upstream, following the sound, till I broke through the undergrowth at a place close to camp, where in the dry season there is a fairly wide sandbank on our side and on the other a dry watercourse which drops abruptly into the river.

Suddenly I stopped unable to believe my eyes.

There was Elsa standing on the sandbank within a few yards of me, one cub close to her, a second cub emerging from the water shaking itself dry and the third one still on the far bank, pacing to and fro and calling piteously. Elsa looked fixedly at me, her expression a mixture of pride and embarrassment.

I remained absolutely still while she gave a gentle moan

to her young, that sounded like *M – hm, M – hm*; then she walked up to the landing cub, licked it affectionately and turned back to the river to go to the youngster who was stranded on the far bank. The two cubs who had come across with her followed her immediately, swimming bravely through the deep water, and soon the family were reunited.

Near to where they landed a fig tree grows out of some rocks, whose grey roots grip the stone like a net; Elsa rested in its shade, her golden coat showing up vividly against the dark green foliage and the silver-grey boulders. At first the cubs hid, but soon their curiosity got the better of their shyness. They began by peeping cautiously at me through the undergrowth and then came out into the open and stared inquisitively.

Elsa *M – hm, M – hm*'d which reassured them and when they were quite at their ease they began to climb on to their mother's back and tried to catch her switching tail. Rolling affectionately over her, exploring the rocks and squeezing their fat little tummies under the roots of the fig tree, they forgot all about me.

After a while Elsa rose and went to the water's edge intending to enter the river again; one cub was close to her and plainly meant to follow her.

Unfortunately, at this moment the Toto, whom I had sent back to fetch Elsa's food, arrived with it. Immediately she flattened her ears and remained immobile until the boy had dropped the meat and gone away. Then she swam quickly across followed by one cub, which, though it kept close to her, seemed to be quite unafraid of the water. When Elsa settled down to her meal, the plucky little fellow turned back and started to swim over on its own to join, or perhaps to help, the other two cubs.

As soon as Elsa saw it swimming out of its depth, she plunged into the river, caught up with it, grabbed its head in her mouth and ducked it so thoroughly that I was quite worried about the little chap.

When she had given it a lesson not to be too venturesome, she retrieved it and brought it, dangling out of her mouth, to our bank.

By this time a second cub plucked up courage and swam across, its tiny head just visible above the rippling water, but the third stayed on the far bank looking frightened.

Elsa came up to me and began rolling on her back and showing her affection for me; it seemed that she wanted to prove to her cubs that I was part of the pride and could be trusted.

Reassured, the two cubs crept cautiously closer and closer, their large expressive eyes watching Elsa's every movement and mine, till they were within three feet of me. I found it difficult to restrain an impulse to lean forward and touch them, but I remembered the warning a zoologist had given me: Never touch cubs unless they take the initiative, and this three-foot limit seemed to be an invisible boundary which they felt that they must not cross.

While all this was happening the third cub kept up a pathetic miaowing from the far bank, appealing for help.

Elsa watched for a time, then she walked to the water's edge, at the point at which the river is narrowest. With the two brave cubs cuddling beside her she called to the timid one to join them. But its only response was to pace nervously up and down; it was too frightened to try to cross.

When Elsa saw it so distressed she went to its rescue accompanied by the two bold ones who seemed to enjoy swimming.

Soon they were all on the opposite side again where they had a wonderful time climbing up the steep bank of a sand lugga, which runs into the river, rolling down it, landing on each other's backs and balancing on the trunk of a fallen doam palm.

Elsa licked them affectionately, talked to them in her soft moaning voice, never let them out of her sight and whenever one ventured too far off for her liking, went after the explorer and brought it back.

I watched them for about an hour and then called Elsa who replied in her usual voice, which was quite different from the one she used when talking to the cubs.

She came down to the water's edge, waited till all her family were at her feet and started to swim across. This time all three cubs came with her.

As soon as they had landed she licked each one in turn and then, instead of charging up to me as she usually does when coming out of the river, she walked up slowly, rubbed herself gently against me, rolled in the sand, licked my face and finally hugged me. I was very much moved by her obvious wish to show her cubs that we were friends. They watched us from a distance, interested, but puzzled and determined to stay out of reach.

Next Elsa and the cubs went to the carcase, which she started eating, while the youngsters licked the skin and tore at it, somersaulted over it and became very excited. It was probably their first encounter with a 'kill.'

The evidence suggested that they were six weeks and two days old. They were in excellent condition and though they still had a bluish film over their eyes they could certainly see perfectly. Their coats had fewer spots than Elsa's or her sisters', and were also much less thick than theirs had been at the same age, but far finer and more shiny. I could not tell their sex, but I noticed immediately that the cub with the lightest coat was much livelier and more inquisitive than the other two and especially devoted to its mother. It always cuddled close up to her, if possible under her chin, and embraced her with its little paws. Elsa was very gentle and patient with her family and allowed them to crawl all over her and chew her ears and tail.

Gradually she moved closer to me and seemed to be inviting me to join in their game. But when I wriggled my fingers in the sand the cubs, though they cocked their round foxy faces, kept their distance.

When it got dark Elsa listened attentively and then took the cubs some yards into the bush. A few moments later I heard the sound of suckling.

I returned to camp and when I arrived it was wonderful to find Elsa and the cubs waiting for me about ten yards from the tent.

I patted her and she licked my hand. Then I called the Toto and together we brought the remains of the carcase up from the river. Elsa watched us and it seemed to me that she was pleased that we were relieving her of the task of pulling the heavy load. But, when we came within twenty yards of her, she suddenly rushed at us with flattened ears. I told the boy to drop the meat and remain still and I began to drag it near to the cubs. When she saw that I was handling the 'kill' alone, Elsa was reassured and as soon as I deposited it she started eating. After watching her for a while, I went to my tent and was surprised to see her following me. She flung herself on the ground and called to the cubs to come and join me. But they remained outside miaowing; soon she went back to them and so did I.

We all sat together on the grass, Elsa leaning against me while she suckled her family.

Suddenly two of the cubs started quarrelling over a teat. Elsa reacted by rolling into a position which gave them better access. In doing so she came to rest against me and hugged me with one paw, including me in her family.

The evening was very peaceful, the moon rose slowly and the doam palms were silhouetted against the light; there was not a sound except for the suckling of the cubs.

So many people had warned me that after Elsa's cubs had been born she would probably turn into a fierce and dangerous mother defending her young, yet here she was trusting and as affectionate as ever, and wanting me to share her happiness. I felt very humble.

4

The Cubs Meet Friends

When I woke up next morning there was no sign of Elsa or the cubs, and as it had rained during the night all spoor had been washed away.

About tea-time she turned up alone, very hungry; I held her meat while she chewed it so as to keep her attention and meanwhile told the Toto to follow her fresh pug marks to get a clue to the present whereabouts of the cubs.

When he returned Elsa hopped on to the roof of my car, and from this platform she watched the two of us walking back along her tracks into the bush.

I did this deliberately to induce her to return to the cubs. When she realized where we were going she promptly followed us, and, taking the lead, trotted quickly along her pug marks; several times she waited till, panting, we caught up with her. I wondered whether at last she meant to take us to her lie-up. When we reached the 'Whuffing Rock,' so named because it was there that we had once surprised her with her mate and had been startled by their alarming whuffing, she stopped, listened, climbed swiftly half-way up the slope, hesitated until I had caught up with her and then rushed ahead till she had reached the saddle of the rock from which the big cleft breaks off on the far side. There, much out of breath, I joined her. I was about to pat her when she flattened her ears, and with an angry snarl gave me a heavy clout. Since it was plain that I was not wanted, I retreated. When I had gone half-way down the face of the rock I looked back and saw Elsa playing with one cub, while another was emerging from the cleft.

I was puzzled at the sudden change in her behaviour, but I respected her wishes and left her and her family alone. I joined the Toto who had waited in the bush just below and we watched Elsa through our field-glasses. As soon as she saw that we were at a safe distance she relaxed and the cubs came out and began playing with her.

One cub was certainly much more attached to her than the others; it often sat between her front paws and rubbed its head against her chin, while the two others busily investigated their surroundings.

George returned on the 4th February and was delighted to hear the good news of the cubs; in the afternoon we walked towards the Whuffing Rock hoping that he too might see them.

On our way we heard the agitated barking of baboons. We thought it very likely that Elsa's presence was the cause of the commotion, so, as we approached the river, we called out to her. She appeared immediately, but though she was very friendly she was obviously upset and rushed nervously backwards and forwards between us and the bush, which grew along the river's edge. She seemed to be doing her best to prevent us from reaching the water.

We assumed that her cubs were there and were surprised that she should try to prevent George from seeing them. In the end she led us back to the camp by a wide detour.

Two days later we saw her near the Whuffing Rock. As we were walking towards it we talked rather loudly to give her notice of our approach. She emerged from the thick undergrowth at the mouth of the cleft and stood very still, gazing at us. After a few moments she sat down facing us – we were still some two hundred yards away – and made it very plain that we were not to come any nearer. Several times she turned her head towards the cleft and listened attentively, but apart from this she remained in her 'guarding' position.

We now realized that she made a difference between bringing the cubs to see us and our visiting them.

Two weeks passed before she brought the cubs to camp to introduce them to George. This was not entirely her fault for during this time we were obliged to go to Isiolo for a couple of days and while we were away she and the cubs had arrived at the camp one morning looking for us, but had only found the boys.

Makedde told us that he had gone to meet her and she had rubbed her head against his legs and one plucky cub had boldly walked up to within a short distance of him.

However, when he squatted and tried to pat it, it had snarled and run off to join the others who were hiding some distance away. They had stayed in camp till lunchtime and then left. Elsa returned alone during the afternoon asking for meat, but the goat carcase was by then very high and she left in disgust after dark.

I arrived about an hour after she had gone. Makedde was delighted with the plucky cub; he said he was sure it was a male and told me he had given it a name, which was, he said, very popular with the Meru tribe. It sounded like Jespah. I asked him and the other boys where the name came from. They said it was out of the Bible, but as each boy pronounced it slightly differently it was difficult for me to trace it. The nearest phonetic association I could find was Japhtah, which means 'God sets free.' If that were the origin of the little cub's name it could not be more appropriate. Later, when we knew that the family consisted of two lions and a lioness, we called Jespah's brother, who was very timid, Gopa, for in Swahili this means timid, and his sister we named Little Elsa.

The next day Elsa arrived in the afternoon; she was extremely pleased to see me and very hungry. After a while I went for a walk hoping that in my absence she would return to her cubs, and when I came back she had gone.

The following morning it was drizzling. I woke up to hear Elsa's typical cub moan coming from across the river; I jumped out of bed and was just in time to see her crossing the river with her cubs, Jespah close to her and the other two some way behind.

She walked slowly up to me, licked me and sat down next to me. Then she called repeatedly to the cubs. Jespah ventured fairly near to me, but the others kept their distance. I collected some meat which Elsa promptly dragged into a nearby bush; she and the cubs spent the next two hours eating it, while I sat on a sandbank watching them.

While they ate Elsa talked continuously to the cubs in a series of low moans. They often suckled, but also chewed at the meat. Elsa did not regurgitate any meat for them, though considering the vast amount that she had eaten lately when she came alone to the camp it seemed likely that she might have regurgitated some of the meat later in the day for the benefit of the cubs. But this is speculation. We never saw her doing it.

The cubs were now about nine weeks old and for the first time I was able to confirm Makedde's belief that

54

Jespah was a lion; he had a brother and a sister and it was now that we gave them their names.

After a while I went off to have breakfast and soon afterwards saw Elsa leading the cubs in a wide circle to the car track. I followed slowly hoping to take some photographs but she stopped suddenly broadside across the road and flattened her ears. I accepted the reproof and went back, turned to have a last look at them and saw the cubs bouncing along behind their mother going in the direction of the Big Rock. By now they were lively walkers, chasing and prodding one another as they tried to keep pace with Elsa. In spite of their high spirits they were most obedient to her call, and were also already well trained in cleanliness and always stepped off the path when they were producing their excrements.

During the next few days Elsa often came alone to visit us. She was always affectionate but some of her habits had altered since she had given birth to the cubs. She now very seldom ambushed us, was less playful, more dignified.

I wondered how she placed her cubs when she came out on these long visits. Did she instruct them not to move till she returned? Did she hide them in a very safe spot?

When, on 19th February, George came 'on duty,' I returned to Isiolo to meet Lord William Percy and his wife and bring them to see Elsa's family.

In general we discouraged visitors, but we made an exception for these old friends who had known Elsa since she was a cub and had always shown the greatest interest in her development.

On our arrival in camp, George greeted us with the news that he had seen the cubs. That morning he woke up while it was still dark and heard quick, short lapping sounds, as well as long laps, coming from the direction of Elsa's water bowl. He looked out and saw the cubs dimly outlined around the bowl; a few minutes later they all went off.

He said that just at the moment in which he had first heard the vibrations of our car, Elsa had been about to cross to our side of the river with her cubs, but when she had become aware of a car approaching she had retired into the bush.

Soon she emerged but seemed nervous and disinclined to enter the water. To induce her to join us I called to her and placed a carcase close to the river.

She made no move till I had gone back to join our friends, then she swam quickly across, seized the goat and rushed back with it to the cubs. Once across she dragged it on to a grassy patch, where the whole family set to and had a good meal; we watched them through our field-glasses.

After it had grown dark we heard fearful growls and by the light of our torches saw Elsa defending her 'kill' from a crocodile, which, when it observed us, disappeared quickly into the water.

In the morning an examination of the spoor showed that in the end the 'croc' had been successful in stealing the carcase. We were impressed by the fact that Elsa always seemed to know just how far she could go with these reptiles. She had never shown any fear of them, although we knew that in this river there were many crocodiles measuring twelve feet or more. She had her favourite crossings and avoided the places where the river was very deep, and besides taking this precaution there can be no doubt that she had some means of sensing the presence of 'crocs.'

How this worked we could not guess. We had our own method of discovering the presence of 'crocs'; we knew that they invariably respond to a certain sound, which can roughly be represented by '*imn, imn, imn*,' and we often took advantage of our knowledge.

If we suspected the presence of crocodiles we would keep ourselves hidden from the river and repeat '*imn, imn, imn*'; then if there were any 'crocs' within 400 yards, they would come to the water's edge as though drawn by a magnet. Often we went on until we saw many ugly peri-scopic nostrils sticking out above the water. If we moved, and our noises then came from a different place, they would follow them.

George had learned this trick from African fishermen on Lake Baringo which is infested by 'crocs.'

Before going knee-deep into the water the fisherman places two men at a little distance on either side of himself. Their job is to attract the 'crocs' by their call and then spear them, while the fisherman works safely in the middle.

We often wondered what '*imn, imn, imn*' conveys to the crocodile. Does it stimulate the mating instinct? Is it reminiscent of the call of their young, and, anyway, how does it reach them? Only their nostrils project above the surface of the water on these occasions; can the sound vibrations reach their ear membranes through the water?

We knew that the 'crocs' responded to a specific sound, what we did not know was whether they themselves emit-ted any vibration which made Elsa aware of their proxim-ity. The facts suggested it, but we had no proof.

Next day, while we were having tea in the studio, Elsa appeared alone; our friends were included in her customary friendly rubbings and she bore with my taking a few photographs, but then walked out of the picture.

She never liked being photographed and since the arrival of the cubs she had become even more averse to it.

Later, Lady William started sketching her and this was another thing she usually disliked, but to-day she seemed to have no objection. All the same, I kept close by in case she might suddenly take a dislike to serving as a model. However, as she appeared quite indifferent to what was going on, after a while I went away. As soon as my back was turned she rushed like lightning at the artist and embraced her playfully. As Elsa weighs about three hundred pounds I admired the calm way in which Lady William accepted the demonstration. After this we decided that the sketching had better cease. Elsa left us at dusk; soon a leopard began coughing and Elsa and her mate started up a lively conversation which lasted throughout most of the night.

At tea-time the next day we saw Elsa and the cubs on the opposite side of the river, but when she spotted us she moved her family a short distance downstream, then they crossed the river. We quickly fetched some meat which Elsa promptly collected and then took into the bush to her cubs who were out of sight.

Later, they all got thirsty and came to the water's edge to drink. I was glad that our guests should have this splendid view of them drinking close together, their heads stretched forward between the pointed elbows of their front legs, which were bent. At first they just lapped noisily, then they plunged into the shallow water and began to play. They were certainly not water-shy, as cats are said to be. A big boulder surrounded by water made a perfect setting for playing 'king of the castle' and I thought of the days when Elsa and her sisters had had to be content with a potato bag on our veranda at Isiolo for their 'castle.'

How lucky these little cubs were to be living in such a lovely and exciting place. The rocky range on which they were born started on our side of the river, crossed it and circled for several miles on the other side. It was broken up by cracks and caves in which hyrax and other small animals had made their homes; around it on all sides stretched the bush, which was full of spoor and of the scent of wild animals, and then there was the river, with its rocks and sandbanks on which turtles, looking like giant pebbles, basked in the morning sun.

Of course there were also the unpleasant and dangerous crocodiles, but they lived mostly in deep pools. These are overhung by doam palms whose feathery fronds sweep into the back waters; the grey-green blotchy skins of the 'crocs' blend to perfection with the decaying vegetation and often make them difficult to detect. In other places the river is bordered by fig trees, acacias and phœnix palms from which lianas and tendrils dangle and twist their way into the thick undergrowth and thereby provide impenetrable hideouts for many animals.

Here live the graceful vervet monkeys, the clowning baboons, the turquoise-coloured agamas, all kinds of lizards, some with bright orange heads, others with vivid blue tails, and also our friend the monitor. Bush buck, lesser kudu and water buck come here to drink and the flattened, trampled ground shows that rhino and buffalo also visit it. Of all the inhabitants of the bush the most fascinating to us are the many coloured birds which throng the bush: the orioles, the brilliant kingfishers, the iridescent sun birds, the fish-eagle and the palm-nut vulture, black and white and very large, the hornbills whose rhythmic croaking rises to a crescendo and only drops to rise again.

It is hard to imagine a lovelier sight than the highly-coloured body of a Hunter's sun bird moving between the glossy foliage and strongly scented white gardenia flowers, which are often as large as saucers.

Among all these birds my special friends were a pair of fin-foots, which I see only occasionally (for they are extremely uncommon), their short red legs racing hurriedly across a sandbank when alarmed.

After our friends had gone to bed, George and I returned to see Elsa. We found her standing at the water's edge facing a crocodile, whose head rose out of the river about four feet away.

We did not want to frighten the cubs by firing a shot, so I tempted Elsa to leave the place by offering her a treat of which she was very fond; it consists of brains, marrow, calcium and cod liver oil. I began giving it to her when she was pregnant and she found it irresistible.

Now she followed the bowl in which I carried it and came with the cubs to sit in front of our tent, facing the bright lamplight.

The cubs were unperturbed by the glare; perhaps they thought it was some new kind of moon.

After I had gone to bed, George turned out 'the moon' and sat for a while in the dark. The cubs came within touching distance of him, then, having had 'a drink for the road,' they all trotted off towards the Big Rock, from which immediately afterwards he heard Elsa's mate calling.

Later George went to collect the remains of the carcase,

but found it had already been pulled into the water by a crocodile. He shot at the thief and rescued the meat.

Early one morning Elsa visited the camp before anyone was up. I heard her and followed her. She was already in the water when I called to her, but she came back at once, settled with me on a sandbank and began to miaow at the cubs, encouraging them to come near us. They approached within three yards but obviously did not wish to be handled, and as the last thing I wanted was that they should become tame, I was very pleased about this.

Elsa seemed puzzled that they should still be scared of me, but in the end she gave up her attempt to make us fraternize, took her family across the river and disappeared into the bush.

At ten o'clock she returned alone, sniffed restlessly in the river bush and then trotted, scenting, along the road she had taken in the morning.

After we had lost sight of her we heard her growling fiercely. She returned along the track still sniffing anxiously and finally roared at full strength towards the rock, after which she rushed into the river and disappeared into the bush on the far side. We did not know to what to attribute her strange behaviour, but thought perhaps she might have lost a cub.

When at lunch-time Ibrahim brought in three tribesmen who said they were looking for a goat which had strayed, but carried bows and poisoned arrows, we felt sure that we had been right: no doubt their arrival had startled the little ones and they had bolted.

Elsa did not bring the cubs into camp again for a couple of days. That morning we had taken our friends to see the magnificent falls of the Tana river, which few Europeans visit because they are so inaccessible.

Here we spent some time watching the hippos wallowing in the shallows and playing affectionately with each other. I thought how unfair it is that because they are so ugly, huge and clumsy one is inclined to be surprised when one sees them displaying the feelings that one would take for granted in a beautiful creature.

At least they have nice voices, whose booming sound recalls the tone of the low notes of a 'cello.

On our return we found Elsa and the cubs in camp and while we had our sundowners they enjoyed their dinner. We were silent for we knew how sensitive the cubs were to the sound of talking. They did not mind the chatter of the boys, far away in the kitchen, but if we were near them and said a word to each other, even in a low voice, they sneaked away. As for the clicking of a camera shutter – it gave them the jitters.

They were ten weeks old and Elsa had begun to wean them. Whenever she thought they had had enough milk she either sat on her teats or jumped on to the roof of the Landrover. So if the cubs did not want to starve they had to eat meat. They tore the intestines of the 'kills' out of their mother's mouth and sucked them in like spaghetti, through closed teeth, pressing out the unwanted contents, just as she did.

That evening one cub was determined to get some more milk and persistently pushed its way under Elsa's belly until she became really angry, gave it a good spank and jumped on to the car.

The little ones resented this very much; they stood on their hind legs resting their forepaws against the car, miaowing up at their mother, but she sat and licked her paws, as though she were quite unaware of the whimpering cubs below.

When they had recovered from their disappointment they bounced off, cheerfully making explorations which took them out of her sight. Elsa became extremely alert if they did not come when she called them, and if they did not reappear quickly she hopped off the car and fetched them back to safety.

The next two evenings Elsa came to camp without her family. She was exuberantly affectionate to all of us and swept the table clear of our sundowners, which made our friends appreciate why in camp we use crockery and glasses made of unbreakable material. On the third evening she brought the cubs with her and behaved in the same way. We were rather surprised to observe that the cubs were not in the least startled when our supper landed on the ground with a noisy clatter.

They now seemed quite at home in our presence, so it astonished us that on the two following evenings Elsa left them at an open salt lick about a hundred yards away, and we were also puzzled to know how she trained them to stay put while she enjoyed a good meal in full view of them.

During all that night it poured without stopping. On such occasions Elsa always takes refuge in George's tent, and now, in she came, calling to the cubs to follow her. But they remained outside apparently enjoying the deluge and soon their poor mother felt it her duty to go out and join them. We heard them playing round the camp and then we thought we heard muffled voices, but the drumming of the rain on the roof of our tent was so loud that it took us a little while before we realized that these were those of our friends. Their tent had collapsed and they were trying to struggle out from beneath the wet canvas.

We went to their help, hoping that Elsa and her cubs would not join the rescue team. Luckily they didn't, and while we hammered in the tent pegs and flashed our torches Elsa stood aside miaowing gently to reassure the cubs. At dawn the rain stopped and she took her family off towards the rock, and we dried our friends' clothes.

Later in the day I went off to Isiolo with them. George stayed on in camp. We knew that, now the rains had started in earnest, transport would soon become very difficult, so we had to make our plans accordingly.

5

The Cubs in Camp

When I came back to camp two days later to relieve George, I found that I had to be careful about letting any of the boys come near Elsa when she had the cubs with her. If even Makedde approached them she flattened her ears and looked at him through half-closed eyes which had a cold, murderous expression. Me, she trusted completely and gave proof of it by sometimes leaving the cubs in my charge when she went to the river to drink.

For several nights we had terrific thunderstorms and the lightning and the crashes came so close together that I was quite frightened. The water poured down as though it were flowing through a pipe.

As George's tent was empty, Elsa and the cubs could very well have sheltered in it, but the youngsters' inbred fear of man was so great that they preferred to soak outside. This trait was the most obvious sign of their wild blood and it was something we were determined to encourage, even at the expense of a wetting and even in defiance of Elsa's wish to make them into friends of ours. Often she seemed to be playing a sort of 'catch as catch can' with them, circling nearer and nearer to the tent in which I was sitting, as though she wanted to bring them into it without their becoming aware of what was happening.

Twice she dashed into the tent and peeping over my

shoulder called to them. But whatever she did they never overstepped their self-imposed frontier.

It seemed that our rearing of their mother in domesticity had in no way impaired the instinct which all wild animals possess and which warns them against approaching an unknown danger. Moreover, Elsa herself had shown by concealing her cubs from us for five or six weeks, that her own instinct for protecting her young was still alive.

Now, she was plainly disappointed that her efforts to make one pride of us were proving unsuccessful, partly owing to the cubs' fear of man and partly owing to what she must have taken as heartless lack of co-operation on our part. She seemed very puzzled, but had no intention of giving up her plan. One evening she entered my tent, deliberately lay down behind me and then called softly to the cubs inviting them to suckle her. By doing this she tried not only to make the cubs come into the tent but also to force them to pass close to me. No doubt they would have been pleased if I had retired behind their mother and she would have been pleased if I had done something to encourage them, but I remained where I was and kept still. To have moved would have defeated Elsa's intention and to have encouraged them would have been against our determination not to tame them. I was sorry because I longed to help the cubs and felt distressed when Elsa looked at me for a long time with a disappointed expression in her eyes and then went out to join her children. Of course she could not understand that my lack of response was due to our wish to preserve the cubs' wild instinct. She plainly thought me unfeeling, whereas I was suppressing all my feelings for the good of her family.

The cubs were worried about our relationship for the

opposite reason and became anxious every evening when Elsa, persecuted by tsetse flies, flung herself in front of me, asking me to dispose of these pests.

When I started squashing the flies and in the process slapping Elsa, the cubs were very upset. Jespah in particular would come close and crouch, ready to spring should his mother be in need of protection. No doubt they found it odd that she should seem grateful for my slappings.

On one occasion when Elsa, Jespah and Little Elsa were drinking in front of the tent Gopa was too nervous to come to the water bowl. Seeing this, Elsa went to him with great deliberation and cuffed him several times, after which he plucked up enough courage to join the others.

Jespah's character was quite different – he was rather too brave. One afternoon after they had all fed and when their bellies were near bursting point Elsa started off towards the rock. By then it was nearly dark. Two cubs followed obediently but Jespah went on gorging. Elsa called twice to him, but he merely listened for a moment and then went on feeding. Finally, his mother came back, and it was in no uncertain manner that she walked up to her son. Jespah realized that he was in for trouble, so gobbling the meat and with large bits of it hanging out of either side of his mouth, trotted after her.

At this time I had to go for a few days to Isiolo while George came to look after the camp.

The way in which the cubs were developing into true wild lions exceeded our hopes, but their father was a great disappointment to us.

No doubt we were partly to blame, for we had interfered with his relationship with his family – but certainly he was of no help as a provider of food for them; on the contrary, he often stole their meat. Moreover, he caused us a lot of trouble. One evening he made a determined attempt to get at a goat which was inside my truck, and another time when Elsa and the cubs were eating outside our tent she suddenly scented him, became very nervous, sniffed repeatedly towards the bush, cut her meal short and hurriedly removed the cubs.

George went out with a torch to find out what the trouble was; he had not gone three yards when he was startled by a fierce growl and saw the cubs' father hiding in a bush just in front of him. He retreated rapidly and luckily so did the lion.

The next day another menace appeared. Makedde reported that an enormous crocodile was sleeping at the place where Elsa usually crossed the river. George took a rifle and went to the spot. The 'croc' was still there, and huge it was, for after he had shot it he measured it – it was twelve feet two inches, a record for that river.

If Elsa had been attacked by such a monster she would not have stood a chance.

Meanwhile, at Isiolo, I was having my own excitements. One afternoon I was sitting on the steps which lead up to the veranda, watching the sun set behind the hills and hoping to see a jackal or some other animal come to our bird bath for a drink. It consisted of a large tyre, cut in half and dug into the ground. By day big birds, squirrels and mongooses came to it and by night water buck, impala, zebra and leopards. I had even seen an elephant empty it at one suck.

During the last few nights I had observed the spoor which showed that a porcupine had come for a drink, and I had placed a banana out for it, hoping to encourage its visits. It was a peaceful and beautiful evening, and I was reluctant to go in, but eventually went to bed. I was alone in the isolated house, except for an armed guard who slept on the back veranda.

Suddenly I was woken by the sound of heavy footsteps; as I listened I heard them go from the sitting-room into George's empty bedroom. Though I had a revolver under my pillow I felt far from brave, but as I did not want to give the burglar time to escape by calling to the guard I got up, with my heart in my mouth, and went very quietly into George's bedroom, expecting to be hit on the head. There was no one in the bedroom and all was quiet, so very cautiously I went towards the bathroom. Suddenly there was a loud rustling behind the door; I was paralysed by fear mainly because I did not associate the noise with anything familiar. Then, before I had time to recover from my fright, a charging porcupine, with upraised quills, rattled threateningly against everything that obstructed its path. I had just time to jump out of its way, then I burst out laughing and in so doing woke up the guard. To my great satisfaction I saw him peeping through the door looking every bit as frightened at the strange noise as I had been. Quite soon the porcupine took itself off, which was considerate of it, as we should have been hard put to it to get rid of our prickly trespasser without injuring ourselves in the process. We remained perplexed as to how he had managed to get up the veranda steps.

I now tried to answer the large mail which had begun to arrive, for Elsa had won the hearts of thousands of people. It was gratifying but we feared she might have to share the fate of all celebrities – lack of privacy.

People from all over the world wrote saying they would

like to come and see her. This was a problem. After all the trouble we had taken to keep her and the cubs wild we could not agree to Elsa and her family being turned into a tourist attraction. We could, of course, appeal to her admirers, to sportsmen and to our friends not to invade her privacy, but we had no legal means of keeping people out and we were very worried in case some visitor should, in our absence, provoke Elsa and accidentally cause trouble.

On my way back to camp I was delayed by floods. We had to cross plains where in some places there was a foot of water and in others deep mud. The car often got stuck and with only Ibrahim and Nuru to help it sometimes took us hours to dig it out and get it going again.

When we reached the fourth and last river which we had to ford, we found it so high that we dared not attempt the crossing. All we could do was to stop and wait until the water level dropped. By this time it was getting dark and the devout Mohammedans faced Mecca, prostrated themselves, and said their prayers.

Nuru had only just returned to us; he had been home for six months because he had suffered from an internal illness. Now he was well again, but he blamed Elsa for his sickness. This surprised me as he had always been very devoted to her, but it seemed that the onset of his malady had coincided with the time at which we had engaged him to look after Elsa and her two sisters. Because of this he was convinced that she had cast 'the evil eye' on him.

It was to dispel this belief that I was now taking him to the camp with me. As we waited in the drizzling rain, I told him about the cubs and he seemed very interested.

During the night the river fell, so we were able to reach

camp in the early hours of the morning. Elsa, attracted by the vibrations of the car, gave us a welcome which, in our exhausted state, we found almost too boisterous.

In the afternoon, hoping to be able to show Nuru the cubs, we all walked in their direction. Suddenly we heard Elsa talking to them in the bush just ahead of us.

Soon she came bouncing out and after greeting us made a great fuss of Nuru. Indeed, she was so overwhelmingly happy to see her old friend again after such a long absence, that he was very much touched; he began to pat her and discarded all his superstitious fear of her 'evil eye.' After this reunion he became even more devoted to her than he had been before his illness. She did not, however, show him her cubs on this occasion and only brought them into camp after dark.

Unlike their mother, they had never had any man-made toys to play with, but they wrestled in the bright lamplight and were never at a loss to find a stick to fight for. At other times they played hide-and-seek and 'ambushes.' Often they would get locked in a clinch, the victim struggling on his back with all four paws in the air. Elsa usually joined in their games; in spite of her great weight, she sprang and hopped about as though she were herself a cub.

We had provided two water bowls for them, a strong aluminium basin and an old steel helmet mounted on a piece of wood, which Elsa had used since her youth. This was the more popular of the two with the cubs. They often tipped it over and were alarmed at the clatter it made when it fell. Then recovering from their fright they faced the shiny moving object with cocked heads and finally began to prod it cautiously. We took flashlight photographs of these games.

We had more difficulty in taking pictures of them at play during daylight, because they were then less active. Our best chance was in the late afternoon, when they went to a favourite playground near to a doam palm which had fallen at the edge of the river-bank, some two hundred yards from the camp. This place afforded all amenities: it overlooked a wide-open space, it had thick bush close by into which they could disappear if any danger threatened, it was near to a salt lick, and also to the river, should they want a drink. Besides this I often placed a carcase nearby.

George and I used to hide in the bush and take films of the family climbing up and down the fallen trunk, teasing their mother who was always there to guard them.

They knew we were near but this did not disturb them; if, however, an African appeared, even in the distance, the game stopped at once and the cubs disappeared into the bush, while Elsa faced the intruder with flattened ears and a threatening expression.

On 2nd April George went back to Isiolo but I stayed on in camp.

As the days passed I observed that the cubs were getting more and more shy even of me. Now they preferred to sneak through the grass in a wide circle to reach their meat, rather than follow their mother in a straight line, because this involved coming very close to me.

To prevent predators from stealing the meat during the night I started dragging the carcase from the doam palm near to my tent, to which I attached it by a chain.

It was often a heavy load and Elsa used to watch me apparently content that I had taken on the laborious task of protecting her meat.

Jespah was much less happy when he saw me handling

the 'kill.' After several half-hearted attacks he sometimes charged me in a proper fashion, first crouching low and then rushing forward at full speed. Elsa came instantly to my rescue: she not only placed herself between her son and me, but gave him a sound and deliberate cuff. Afterwards she sat with me in the tent for a long time, totally ignoring Jespah, who rested outside looking bewildered. He lay by the helmet bowl, his head against it, occasionally lapping lazily.

Touched as I was by Elsa's reaction, I also understood that Jespah should be disconcerted by his mother's disapproval of his instinctive reaction and I was most anxious not to arouse his jealousy.

He was still too small to do very much harm but we both recognized that it was essential to establish a friendly truce with the cubs while they were still dependent upon us for food and before they had grown big enough to be dangerous. It was a difficult problem because while we did not want them to be hostile, neither did we want them to become tame. Recently Elsa herself seemed to have become aware of our difficulty and to be making her contribution to solving it. While she spanked Jespah if in his attempts to protect her he attacked me, she also dealt firmly with me if she thought I was getting too familiar with her children. For instance, several times when I came close to them while they were at play, she looked at me through half-closed eyes, walked slowly but purposefully up to me, and gripped me round the knees in a friendly but determined manner, which indicated very plainly that her grip would become much firmer if I did not take the hint and retire.

Elsa spent much of the time on the Big Rock and one

afternoon, taking Nuru with me to help carry the heavy paraphernalia, I went there with my cameras, hoping to make pictures of the lions silhouetted against the sky or, a scene I particularly liked, Elsa coming down the steep rock, followed by her family and waiting for the cubs as they negotiated the more difficult passages. Just as we had set up the cameras she appeared on the top of the rock; as soon as she spotted us she sat down and did not move for an hour; then she disappeared. Before sunset she returned, but seeing that we were still there retired. Only after it was too dark to take any photographs and we were packing up did she reappear with her cubs. She had not been to camp for two days and I thought she must be very hungry. I wondered whether her absence was due to her being with her lion, or whether she objected to the boys.

6

The Personality of the Cubs

One morning I was woken up by the arrival of a Landrover bearing a message which told me to expect the arrival of Godfrey Winn and another journalist, Donald Wise. They were to land at the nearest airstrip and there George was to meet them and bring them with their pilot to camp for the night.

I was worried by this at first, because Elsa's reactions were unpredictable when she had the cubs with her – she had lately even objected to Nuru's presence. What would she make of strangers? I sent the driver back with a message begging George to halt the party ten miles away from the camp and suggested that I would meet them there. In order to be sure that they paid attention to my request, after lunch I sent Ibrahim a few miles along the road to intercept and stop the car if he should sight it. He was the bearer of a second note.

Having taken all these precautions I was rather surprised when the party nevertheless turned up, and I was trying to argue our guests into retiring when I heard Elsa's '*mhn, mhn*.' Probably she had been attracted by the vibration of the engine; anyway, there she was and the cubs with her. In the circumstances all I could do was to make the best of the situation.

While our guests were settling into the camp George

tied a carcase to the fallen doam palm trunk, so that we could watch Elsa and the cubs eating. I told Mr Winn that I had no wish to monopolise Elsa and her family but was anxious that the lions should live a wild life, which entailed preserving their privacy, and also that I wanted to avoid any possible future trouble, which might arise from their getting too used to visitors. He and Mr Wise were most understanding and promised me that whatever they published about this visit would lead readers to respect Elsa's wild life, which we had only managed to establish at the cost of great patience and sacrifice.

We spent a pleasant evening together, dining beside the tent. After a while Elsa jumped up on the Landrover only a few yards from us: later she allowed Mr Winn to stroke her. When they left the next morning I felt very sorry that our guests had at first met with such a bad reception.

On the following evening we tied up a carcase near our tent. Elsa soon came for her meal and did all she could to induce the cubs to join her. She pranced round and did her best to cajole them and tried by every means to break down their fear, but not even Jespah ventured into the lamplight. That evening we heard their father calling and by the next morning they had all gone.

When, on the 8th April, George left for Isiolo I stayed on. One night Elsa turned up her nose at the meat I offered her; afterwards the boys told me that that goat had been ill; so her instinct had evidently warned her that the meat was infected. The cubs also would not touch it. As a rule, they were remarkably greedy, ate enormously and insisted on being suckled by Elsa as well as eating meat.

Elsa spent that evening resting her head against my shoulder and 'mhn-mhning' to the cubs, a very sonorous

The third cub kept up a pathetic miaowing
from the far bank

She brought it, dangling out of her mouth,
to our bank

I noticed immediately that the cub with the lightest
coat always cuddled up under her chin if possible

A moment later a cow buffalo charged Elsa

The monitor was very large, about five feet long,
and blown up to capacity

A dead crocodile: the end of one potential menace

An afternoon visit from the cubs at four months

Visitors' book – spoor in the sand

The cubs were growing more shy every day and preferred to eat outside the area lit by our lamp

Elsa came to our tent at dusk so that I could
rid her of the tsetse flies

Interruption

Each cub receives Elsa's affection in turn

Elsa and Jespah

sound, although it came through closed lips; fruitlessly, she tried to make them come to me.

I was always touched by the way in which she discriminated when she played with me or with them. With the cubs she was often rather rough, pulling their skin, biting them affectionately or holding their heads down so that they should not interfere with her meal; it would have been most painful if she had treated me in the same way, but she was always gentle when we played together. I attribute this partly to the fact that when I stroke her, I always do so very gently, talking to her at the same time in a low, calm voice, to which she responds quietly. I am sure that if I treated her roughly, it would provoke her to demonstrate her superior strength.

That night, after I had gone to bed, I heard Elsa's mate calling, but instead of going to him, she tried to creep through the thorn fence into my boma. I called out, 'No, Elsa, no,' and she stopped at once. She then settled her cubs by the wicker gate and there they spent the night.

The next day she did not appear till after dark and then only brought two cubs with her. Jespah was missing. Elsa settled down to her meal with Gopa and Little Elsa. I was anxious about Jespah but in the dark I could not go and look for him, so I tried to induce his mother to do so, by imitating his high-pitched 'tciang-tciang' at the same time pointing to the bush. After a while, she went off. The two cubs did not seem to be worried by her absence and went on eating for at least five minutes before they made up their minds to follow her. A little later the three of them returned, but there was still no sign of Jespah. I repeated my tactics and Elsa made another search but again returned without him; a third time I induced her to go to find him but this proved equally unsuccessful.

I then discovered that Elsa had a large thorn stuck deeply into her tail. It must have been very painful, and when I tried to pull it out she became irritable. Luckily, I did eventually manage to extract it, then she licked the wound and afterwards my hand, by way of thanking me. By this time Jespah had been missing for one hour.

Suddenly and without any prompting from me she and the two cubs walked purposefully off into the bush and soon I heard Jespah's familiar 'tciangs.'

Presently he appeared with the others, nibbled at some meat and came to lie within five feet of me. I was thankful to see him safely back as the hour he had chosen to go off on his own was the most dangerous so far as predators are concerned, and he was still much too young to tackle even a hyena let alone a lion. I suspected that he had been at the diseased carcase which his mother had refused to touch and which I had ordered to be thrown away at a good distance from the camp.

To provide him with something harmless on which to spend his energy I got an old inner tube and wriggled it near him. He attacked it at once and soon his brother and sister joined in the new game. They fought and pulled until there was nothing left but shreds of rubber.

That night it rained. In the morning I was much surprised to see not only Elsa's pug marks, but those of a cub inside George's empty tent. It was the first time that one had entered the self-imposed forbidden area.

On the following night Elsa, observing that the boys had forgotten to place thorn branches in front of the entrance to my enclosure, pushed the wicker gate aside, entered the tent and promptly lay down on my bed. Wrapped up in the torn mosquito netting she looked so content that I saw myself having to spend the night sitting in the open.

Jespah followed his mother into the tent and stood on his hind legs examining the bed, but fortunately decided against trying it out. The other cubs stayed outside.

We spent most of the evening trying to lure Elsa out of my tent – it was a difficult task since we dared not open the door in case all the cubs were to rush in and join their mother. What we intended was that Elsa should crawl out through the wickerwork door. For some time our hopes of success were pretty dim, then I began to make '*tcianging*' noises round the camp and to flash my torch, pretending that the cubs were lost and that I was looking for them. This soon caused both Elsa and Jespah to rush out. She came through the door, how he got out I do not know. I now had my tent to myself but was unable to sleep because Elsa noisily attacked my truck. However, as on a previous occasion, to my surprise, she stopped when I shouted, 'No, Elsa, no,' to her. I could not understand why she went for the goats' truck, for if she were hungry there was still some meat down by the river.

The cubs were about sixteen weeks old and by now the family should have been guarding its kill. Had Elsa become so lazy that she expected us not only to provide her with food but also to relieve her of the task of protecting it?

Were we ruining her wild instincts and should we leave her? The moment did not seem a propitious one for deserting her, because we had recently found the footprints of two strange Africans very near the camp. No doubt they had been reconnoitring our whereabouts, for the drought was again with us and probably they intended to bring their stock into the game reserves to graze, though this was illegal. In the circumstances, I felt I must go on providing the family with food; if not, Elsa would surely kill some trespassing goat. I comforted myself with the thought that very soon the rains would come, the tribesmen would go away and by the next dry season Elsa would have the cubs well on the run to hunt with her.

Meanwhile, I was immensely interested in observing their development. Already they stretched their tendons; they stood on their hind legs and dug their claws into the rough bark of certain trees – preferably acacias – in so doing they exposed the pink bases of their claws. When they had finished this exercise, the bark showed deep gashes.

I noticed a curious fact about Elsa's fæces, which I had previously often examined for parasites. Before she gave birth to the cubs I had always found them riddled with tapeworm and round-worm, and although I had been told that the presence of tapeworm in a lion's intestines is beneficial (and indeed in the post-mortems we made of any lion George had been asked to shoot, we always found quantities of them), I had nevertheless dosed Elsa from time to time to keep her clear of worms. But since she had had her family I never found a trace of a worm in her fæces nor were there any in those of the cubs. Only after they were nine and a half months old did I find tapeworms in all their droppings again.

Another change related to cleanliness. In the past she had often wetted the groundsheet inside the tent and even sometimes the canvas roof of the Landrover, but since she had become a mother she never permitted herself such bad manners and made the cubs walk off the path whenever they needed to relieve themselves.

None of them showed any sign of the 'ridge back' which is so characteristic of lions. It is a patch about one foot long and two or three inches wide down the middle of the spine on which the hair grows in the opposite direction to the rest of the coat. Elsa and her sister, the Big One, grew their ridge backs very early, but Lustica, the third sister, never developed one.

The cubs were very easily distinguishable. Jespah was much the lightest in colour, his body was perfectly proportioned and he had a very pointed nose and eyes so acutely slanted that they gave a slightly Mongolian cast to his sensitive face. His character was not only the most nonchalant, daring and inquisitive, but also the most affectionate. When he was not cuddling up against his mother and clasping her with his paws he demonstrated his affection to his brother and sister.

When Elsa ate I often saw him pretending to eat too, but in fact only rubbing himself against her. He followed her everywhere like a shadow. His timid brother Gopa was also most attractive; he had very dark markings on his forehead but his eyes, instead of being bright and open like Jespah's, were rather clouded and squinted a little. He was bigger and more heavily built than his brother and so pot-bellied that at one time I even feared he might have a rupture. Though he was by no means stupid, he took a long time to make up his mind and, unlike Jespah, was not venturesome; indeed, he always stayed behind till he was satisfied that all was safe.

Little Elsa fitted her name, for she was a replica of her mother at the same age. She had the same expression, the same markings, the same slender build. Her behaviour, too, was so strikingly like Elsa's that we could only hope that she would develop the same lovable character.

She knew of course that for the moment she was at a disadvantage compared to her two stronger brothers, but she used cunning to restore the balance. Though all the cubs were well disciplined and obeyed Elsa instantly on all important occasions, when playing they showed no fear of her and were only occasionally intimidated by the cuffs she gave them when they became too cheeky.

One evening when the whole family were lying in front of the tent, I started to light the Tilley pressure lamp. Suddenly it burst into flames and I had only time to throw it on to the ground outside the tent before it flared up so alarmingly that I ran for Ibrahim to help me put it out. We collected some old rags to beat it with but by the time we returned it had gone out. During all this commotion the cubs lay very close, quietly watching the strange behaviour of their 'moon.' Elsa also came up to investigate the blaze and I had to shout, 'No, Elsa,' in my most commanding voice to prevent her from singeing her whiskers. She and the cubs then settled outside my tent for the night.

Before I went to sleep I heard what sounded to me like the love-making of a pair of rhinos. These bulky beasts utter the most unexpectedly meek sounds when mating. Another possibility was that the noises came from a pair of buffalo. But whatever it was I was glad that my rifle was near to my bed in case of an emergency. However, nothing more happened and I went to sleep, to be woken up next morning by the sound of crockery clattering on to the ground. The next moment the Toto rushed into the tent minus the tea tray. Breathlessly, he told me that as he was carrying my early morning tea into the tent he had been nearly knocked down by a buffalo. He had only just managed to reach the gate of my enclosure ahead of the beast and to close it in his face. It made me smile to think that a light wicker gate should have given the poor fellow a sense of security when pursued by a charging buffalo. But fortunately the beast must also have been impressed by the flimsy framework and retired.

A more efficient lesson in security measures had recently been given us by a hornbill.

Readers of *Born Free* will remember how Elsa once saved us from walking head-on into a spitting cobra which was coiled up at eye level in the fork of a nearby tree. Since then we had discovered that this snake had made its home in the tree, in a hole about two feet above the ground. We had often watched it and once when we were searching for Elsa's cubs the cobra reared up within a few feet of George, who shot at it but missed. Recently we had noticed that this hole had been sealed, with the exception of a narrow slit, by a mixture of earth, fibre and probably saliva which produced a cement-hard substance.

Through the slit we observed something moving. We were not keen on investigating the hole as we supposed that it had become a cobra nursery. The next thing we noticed was that there were often fresh bird droppings at the base of the tree. Finally, we identified the inhabitant as a young hornbill.

We knew that the female hornbill is imprisoned during the time in which the eggs are hatching, leaving only an opening big enough to enable the male to feed his family. All the same, we were surprised that this pair should have made their nest so close to the ground and also that they had chosen a home which belonged to a cobra. For that matter what had the cobra thought of the proceeding? Perhaps he, too, had just had a family elsewhere, for in the branches of a tree, some fifteen yards away from the hornbill's home, I had recently seen a young bright terra-cotta-coloured cobra. At our approach he wriggled smoothly along the branches and into a hole in the trunk. Both these trees were of the commiphora species. The incident seemed to prove that cobras are good climbers and that they stick to one area. The hornbills remained for six or seven weeks in their home, then we found the sealing substance scattered on the ground and the hole returned to its normal size.

By the time the cubs were eighteen weeks old Elsa seemed to have become resigned to the fact that their relationship with us would never be the same as ours with her.

Indeed, they were growing more shy every day and preferred to eat outside the area lit by our lamp, except for Jespah, who, as he followed his mother everywhere, often came with her into the 'danger zone.' Elsa now often placed herself between us and the cubs in a defensive position.

As they were in excellent condition we thought that we should risk leaving them to hunt with Elsa, anyway for a few days. Their father had been about lately and as the family had only come into camp for short feeding visits, we assumed that they were spending most of their time with him.

While the boys were breaking camp I went to the studio, and sitting on the ground, with my back against a tree, started reading a huge bundle of letters from readers of *Born Free*. They had come up with the Landrover which had arrived to transport our belongings. I was worrying about how I should find time to answer them all, as I wanted to, when suddenly I was squashed by Elsa. As I struggled to free myself from beneath her three hundred pounds the letters were scattered all round the place and, when I had got on to my feet again and begun to collect them, Elsa bounced on to me every time I bent down to pick one up and we rolled together on the ground. The cubs thought this splendid fun and dashed round after the fluttering paper. I thought that Elsa's admirers would have enjoyed seeing how much their letters were appreciated. In the end, I am glad to say that I recovered every one of them; I sent for Elsa's dinner and this diverted her attention and that of the cubs.

By this time the boys had finished packing and the loaded cars were waiting some distance away.

In spite of the loud noise of the cataracts Elsa at once heard the vibrations of the engines. She listened alertly and then looked up at me, her pupils widely dilated, so that her eyes seemed almost black. I had a strong impression that as on previous occasions she realized we were about to desert her and her expression seemed to say: 'What do you mean by leaving me and my cubs without food?' Then she abandoned her half-eaten meal, moved slowly down the sandy lugga with her children and disappeared.

7

Elsa Meets her Publisher

After a five-day absence we returned on the 28th April to camp; ten minutes later Elsa arrived alone. She was in excellent condition and delighted to see us, but made away with the carcase we had brought for her before we had time to tie it up for the night.

She did not reappear for twenty-four hours, then she came alone, ate enormously and by the morning was gone.

The absence of the cubs worried us, the more so because Elsa's teats were heavy with milk, but to our relief the next afternoon we found the whole family playing in a dry river-bed. They followed us back to the camp. Soon afterwards a thunderstorm broke out, Elsa at once joined us in our tent, but the cubs sat outside, at intervals shaking the water off their coats. No one looks his best when drenched and cold, but the cubs certainly looked most endearing, if rather pathetic: their ears and paws seemed twice their normal size against their soaking bodies. As soon as the worst of the downpour was over Elsa joined them and they had an energetic game together, perhaps to warm themselves. After this they settled down to their dinner and tore at the meat so fiercely that beneath their coats which now were dry and fluffy we could see the play of their well-developed muscles. At the end of their meal we, for the first time, saw them bury the uneaten part of the 'kill.' They scratched sand over the little pile most carefully until nothing of it could be seen. Perhaps their mother had taught them to do this during the five days in which they had lived totally 'wild.' After everything had been neatly cleaned up the cubs settled round Elsa and she suckled them for a long time.

As this visit of ours was intended to be a short one, we were anxious to take some photographs but Elsa defeated all our efforts by spending most of her time away from camp. We also wanted to feed her up before another absence, so early one morning we called to her from the foot of the Big Rock. She came down with Jespah at her heels. The other two kept at a little distance. For a time they followed us along the car track, the cubs gambolling and wrestling and Elsa often pausing to wait for them. It was a glorious morning, the air still brisk and the beautiful clouds which usually pattern the Kenya sky on even the brightest days had not yet had time to form. Full of *joie de vivre* the cubs bustled along, knocking each other over, until Elsa turned into the bush, probably intending to take a short-cut to the camp. Little Elsa and Gopa chased after her, but Jespah stayed on the track. It seemed that he felt in charge of his pride and we were certainly not included; he was making sure that we were not following. He paid no attention to his mother's call, and advanced towards us in a most determined fashion, sometimes crouching low and then making a short rush forward. When he was quite close he stopped, looked at us and rolled his head from side to side. He appeared embarrassed and as though he did not know what he should do next. Meanwhile, Elsa returned to fetch her disobedient son, who, having stepped nimbly aside to avoid a vigorous cuff, trotted off after his brother and sister.

We spent a happy day in the studio where the family gorged on a carcase. When they could eat no more, the cubs rolled on their backs and with paws in the air dozed off. I leant against Elsa's stern and Jespah rested under her chin. As soon as the cubs recovered from their siesta they explored the low branches which overhang the rapids half-way across the river. They seemed to have no fear of heights or of the rushing water below and turned with the greatest ease on even the thinnest boughs. Suddenly Elsa stiffened and listened intently, then quickly got up and took the cubs into cover. The next thing I heard was the rumble of approaching elephants. I followed the sound and saw four elephants coming down to the river-bank on the far side, for a drink. We were down wind and they were unaware of our presence. After they had satisfied their thirst they moved slowly back into the bush and disappeared.

When it was nearly dark I began to drag the remains of the meat back to camp. While I was doing this Jespah charged at me twice, but Elsa gave him such a disapproving look that he stopped and sneaked away.

In the afternoon of a day on which George had to go off on patrol, I made another attempt to get some photographs. I took the Toto with me to help carry the cameras and found the family, all very sleepy, in what we call 'the kitchen lugga,' a sandy part of a dry river-bed. When I had spotted them I told the Toto to return to camp. It was very hot, but the sky was overcast and there were some dark rain clouds. I placed the cameras in position and Elsa came up and rolled between the tripods, but without upsetting them. The cubs appeared and were much intrigued by the shining objects and anxious to investigate the bags which I had hung out of their reach. Soon it began to drizzle, but, as it was the kind of shower that never lasts long, I slipped plastic bags over the cameras and did not bother to move them.

Suddenly I saw Elsa standing rigid and looking through half-closed eyes in the direction I had come from.

Then with flattened ears she rushed into the bush like a streak of lightning. I heard a yell from the Toto and dashed after her shouting, 'No, Elsa, no.' Luckily, I was in time to control her. I called to the Toto to make his way back to the camp very slowly and quietly, so as to give Elsa no incentive to chase him. I realized that, seeing the rain, he had decided, against my orders, to come back and help me move the heavy cameras. He narrowly missed being very ill-rewarded for his kindness.

As soon as he was out of sight I succeeded in calming Elsa, by stroking her and telling her over and over again in a reassuring voice that it was only the Toto, Toto, Toto, whom she knew so well. Then I packed up the equipment and started back to the camp. It was not an easy return. Elsa remained very suspicious, she kept rushing ahead of me to make sure that all was safe. As a result I often found myself between her and the cubs and this they did not like. Jespah kept charging me. Eventually I managed to lead the party, which was my intention, for I did not want Elsa to be the first to arrive in camp. I was handicapped because in order to see what was going on behind me I was obliged to walk backwards, carrying my heavy load and constantly talking to Elsa in a casual, reassuring tone of voice, hoping to get her into a peaceful frame of mind before we reached home.

When I was within earshot of the boys I shouted to them to provide a carcase, and I kept Elsa back until it was in position. As a result our return went off peacefully.

After George came back we made another photographic expedition. We went close to the rock where in the morning we had seen Elsa but though we called to her she did not appear. Only after the light had become too weak for filming did she suddenly emerge, silently from a bush only ten yards away from us.

She seemed very composed, perhaps she had spent all the afternoon there watching us. She rubbed her head against our knees but made no sound. We knew she kept silent when she did not wish the cubs to follow her. As quietly as she had appeared she vanished into the bush. Later we saw the pug marks of her lion and concluded that they must be together.

The next afternoon I saw Elsa through my field-glasses near to the spot where she had disappeared on the previous afternoon. She was on the ridge outlined against the sky, watching intently a little gap between some rocks. Though she saw me, she paid no attention to me. I remained there till it was nearly dark, and during all that time she never moved, and seemed to be on guard. Then suddenly her attention became fixed in the direction of the track; probably she heard the sound of George's car returning from patrol.

Soon it appeared, stopped and I got into it and began talking to George. In the back I observed some guinea fowl which he had shot, and looked forward to a pleasant change from the tinned food on which we had been living.

But with a rush Elsa had leapt between us and was among the birds. Feathers began to fly in all directions as she jumped about making frantic efforts to pluck the birds. It looked as though nothing would be left of them, so George picked up a guinea fowl and threw it to the cubs. Immediately Elsa rushed after it and we took the opportunity to start up the engine and move off. Seeing this, Elsa bounded on to the roof of the Landrover and insisted on being driven home. We hoped that after we had gone a few hundred yards her motherly instinct would make her return to the cubs, but she felt far from motherly and we had to bang from the inside on the canvas roof until we made her quite uncomfortable before she decided to jump off and rejoin her bewildered family.

Later they all came to camp and had great fun with the guinea fowl. We were amused to observe how very cunning Little Elsa had become. She allowed her brothers to pull out the prickly quills of the feathers and then when the bird had been nicely plucked took the first opportunity of grabbing it.

After this she defended it with snarls, growls and scratchings, her ears flattened and such a forbidding expression that the boys thought it wiser to go off and pluck another bird. Sometimes the fights between the cubs over food were quite rough, but they never sulked afterwards or showed any resentment. We were surprised that they preferred guinea fowl to goat meat. When she was a cub Elsa had regarded a dead guinea fowl merely as a toy and seldom considered eating it.

The family spent that night close to the camp and in the morning we thought we knew why, for father's pug marks were all around the place and we assumed that he had intended to share their meal. Elsa had obviously not been agreeable to this plan for she had dragged the carcase into a thicket between our tents and the river, where it was unlikely that he would care to come.

She remained with her cubs in this stronghold for the next twenty-four hours, and only left it when she heard George returning from patrol in his Landrover. He had brought some more guinea fowl and the fun and feast of the night before were repeated.

At dusk I went for a stroll and was surprised to see the pug marks of Elsa's lion superimposed on the tyre marks of George's car which had just returned. Father must have been around very recently. When I got back I found Elsa listening very attentively and soon afterwards she moved the cubs and the carcase into her stronghold. A few moments later we heard the lion 'whuffing' close by; he went on all night.

The next morning we had to return to Isiolo for eight days. Though Elsa must certainly have heard the familiar noises of breaking camp, she never emerged from her thorny fortress.

On our return to Isiolo we were thrilled to hear that a call from London had come through three times in the last few days and was now booked for the next morning.

To speak to someone in England, four thousand miles away, is very exciting when one is in a remote outpost. The voice we heard was that of Billy Collins accepting our invitation to come out and meet Elsa. For his arrival we fixed a day during the following week; this would make it possible for him to be with us on our next visit to Elsa.

We chartered a plane to bring him from Nairobi to the nearest place at which an aeroplane can land and then, two days beforehand, we set off. We were determined to find Elsa and try to keep her and the cubs near to the camp so that she should be there to meet her publisher.

Our journey back proved a troublesome one; we had several punctures and, in the end, were obliged to camp in open bush country at a spot where there had recently been a grass fire. Everything around us was black and fire ash drifted in everywhere. We had taken a small supply of drinking water and wished we had a lot more so that we could wash, for we soon became as black as chimney sweeps. Next morning we arrived in camp. George fired a shot to notify Elsa of the fact, and soon we heard her 'hnk-hnk' but she did not turn up. As her voice came from the direction of the studio, I went to it and saw her and the cubs by the river drinking. She glanced at me and went on lapping, as though she were not in the least surprised to see me after eight days' absence.

But later she came up and licked me, and Jespah settled himself about a foot away; then she sprang on to the table and lay stretched at full length on it. Jespah stood on his hind legs and rubbed noses with her. Though they ate a little of the meat I had brought them, they did not seem hungry. However, when George tried to rescue the remains of the carcase, Elsa pulled it gently away from him and took it into a thicket. During the evening we heard Elsa's mate calling and around midnight George woke up to find her sitting on his bed and licking him, while the cubs sat outside the tent watching her.

In the morning I set off with Ibrahim, Makedde and the cook to meet Billy Collins. We took camping equipment with us, for we were not sure when he would arrive and had to provide for the possibility of having to spend a night in the bush on our return journey. When we passed the Big Rock I saw Elsa outlined on its top, watching us drive away. After we had gone about five miles we met a herd of some thirty elephants, which had several young calves with them. Luckily they had crossed the car track just before we arrived and were moving steadily away from us.

We had left very early and in consequence saw an unusual amount of game, bush buck, zebra, water buck, gerenuk and wart-hogs which kept to the bush while herds of Grant gazelles, impala and eland grazed on the open plain.

We were not surprised to see the eland as they always keep to the same area and we knew this herd well. There were ostriches, too, and enormous flocks of guinea fowl chasing each other about over the lava and looking like rolling stones. Most amusing were the baboons, standing up like ninepins in the long grass to get a better view of us. I could only wish that all these lovely animals would be there when we came back so that our guest should see them, though I hoped I might be spared introducing him to the elephants, at least until George was with us.

At lunch-time we arrived in the little Somali village where we expected the aeroplane to land, and I told the Africans to keep the airstrip free of livestock, as a plane might arrive at any moment.

This airfield was originally made for locust control; only a few bushes needed to be cleared to bring it into existence. It is now seldom used and, as the local herds often cross it, blends so well into the surroundings that it is difficult to find from the air.

About tea-time we heard the vibrations of an engine,

but it was a long time before the circling aircraft landed. Then the airstrip was suddenly covered by the entire village population, chattering excitedly. The colourful turbaned Mohammedans, clad in loose-falling garments, watched Billy Collins and the pilot clamber out from the small cabin. Billy had only arrived three hours earlier at Nairobi after a night flight in a Comet. I thought it very sporting of him to venture immediately afterwards on this rather different flight in a four-seater, bumping through notorious air-pockets round the massive Mount Kenya and searching for the small airstrip in the vast sandy plains of the Northern Frontier. The shiny roofs of the few mud-walled native shops had not been of much help in guiding the pilot and that he detected the place was due rather to the crowd of camels and donkeys round the airstrip than to the limp wind-sock and the white corner stones. I could sympathise with the pilot who was anxious to return at once so as to find his way home before dark over the sea of sand and bush below. As we also had a long and rough trip ahead of us, we stopped only for a quick tea at the Government guesthouse and soon set off.

I knew that Billy was very fond of animals, but how far this might allow for safaris under improvised conditions I had no idea. I was worried when he told me that his only camping experience so far had been at a comfortable resthouse on a South Sea island, but when I saw that in spite of the jolting and bumping he was fascinated by every bird, plant or animal, I was reassured. We drove on until dark, then we stopped for a gin at one of the four rivers we had to cross. Expecting Billy to be tired after his long flight from London and also feeling not too happy about the possibility of meeting elephants in the dark, I suggested camping there for the night, but after a discussion with Ibrahim and the Game Scout we decided to drive on.

When we reached the outpost where Elsa's goat deposit is stationed the man in charge gave me a note for George, asking urgently for his presence next day at the nearest administration post as witness in a game case. After two more hours of brushing and winding our car through thick bush, we arrived at camp, ready for a reviving drink, but before George had time to pour it out we heard the familiar 'hnk-hnk' and a few moments later Elsa came rushing along, followed by her cubs. She welcomed us in her usual friendly manner and after a few cautious sniffs also rubbed her head against Billy, while the cubs watched from a short distance. Then she took the meat and dragged it out of the lamplight into the dark near my tent, where she settled with her children for their meal. While this went on we had our supper. We had made a special thorn enclosure next to George's tent for Billy's tent and after introducing him to his home, barricaded his wicker gate from outside with thorns and left him to a well-deserved night's sleep.

Elsa remained outside my tent enclosure and I heard her softly talking to her cubs, until I fell asleep. At dawn I was woken by noises from Billy's tent and recognized his voice and George's: evidently they were trying to persuade Elsa to leave Billy's bed. As soon as it got light she had squeezed herself through the densely woven wicker gate and hopped on to Billy's bed, caressing him affectionately through the torn mosquito net and holding him prisoner under her heavy body. Billy kept admirably calm considering that it was his first experience of waking up with a fully-grown lioness resting on him. Even when Elsa nibbled him slightly in his arm, her way of showing her affection, he did nothing but talk quietly to her.

Soon she lost interest and followed George out of the enclosure where she romped round the tents with her cubs as if her visit to Billy had only been a morning call on the new friend. Afterwards the family disappeared towards the Big Rock and later George left to attend the court. Billy and I spent the day in the studio discussing publication problems until George returned at teatime.

He told us that he had just passed a herd of elephant close to camp, so we finished our tea quickly and drove along the track to film them, but when we came to the Big Rock we noticed Elsa on its top posing magnificently against the sky. We forgot about the elephants and walked to the base of the rock, hoping to film Elsa and her cubs. As she repeatedly listened to some sound coming from behind a large boulder nearby it seemed likely that they were close. Elsa watched our every step, but never moved, however coaxingly we called to her. She kept aloof, and the cubs did not appear. We waited for a considerable time but as nothing happened we decided to try our luck with the elephants.

As soon as we had returned to the car Elsa stood up and called her cubs; as if to tease us, all of them now posed splendidly. We had been waiting for over one hour for just this. However, as Elsa had made it so clear that she was in no mood to be filmed, we drove on to the spot where George had met the elephants, but we found nothing but their footmarks and we returned to Elsa.

By the time we reached the rock the light was too weak for photographing, so we just watched the family through our field-glasses. The cubs chased and ambushed each other round the boulders while Elsa kept her eyes fixed on us. Finally, we called her and she came down at once, rushed through the bush and, after greeting us all affectionately, landed with a heavy thud on the roof of the Landrover. While we patted her paws which dangled over the windscreen, she watched the cubs which were still playing on the rock quite unconcerned at her departure. Though Elsa seemed to enjoy our attentions, she never took her eyes off her children until they finally scrambled down the rock. Then she jumped off the car and disappeared into the bush to meet them.

We took this opportunity to drive home and prepare a carcase for the family. As soon as it was ready they arrived and began to tear at the meat, while we had our sundowners a few feet away. All that evening we watched the lions who seemed to have accepted Billy as a friend.

Before daybreak I was again woken up by noises coming from his tent, into which Elsa had once more found her way to say good morning. After some coaxing from George, who had come to his rescue, she left. George then reinforced the thorns outside the wicker gate with such a bulk that he felt sure Elsa would not be able to penetrate this barricade, so he went to bed again. Unfortunately, during the night George had developed a go of fever and felt far from well. These sudden attacks of malaria sometimes come on in spite of prophylactic treatment in people like George whose organism has been weakened by taking anti-malarial drugs for over thirty years. But Elsa was not going to be defeated by a few thorns and so after a short time Billy found himself again being embraced by her and squashed under her weight. While he struggled to free himself from the entangling mosquito net George came to his rescue, but this time he took much longer to remove the thorns outside the gate, and by the time he got inside Elsa had managed to clasp her paws around Billy's neck and held his cheekbones between her teeth. We had often watched her doing this to her cubs; it was a sign of affection, but the effect on Billy must have been very different. It was very remarkable that he did not lose his head. By the time I arrived Elsa had left the tent and was playing with her cubs near the river bush. I investigated the slight scratches Elsa had left on Billy's shoulder, but luckily they were all superficial and with a dressing of M. and B. powder healed within two days.

I was very much alarmed at Elsa's unusual behaviour. She had never done anything like this to a visitor and I could only interpret it as a sign of affection; if she had not done it in play she could have acted in a very different way, but whatever her motive may have been, I was very upset and remained with Billy in his tent until Elsa, I hoped, had taken her cubs away for the day. In spite of my precaution she forced herself a third time through the wicker gate before either George who was outside or I who was inside could stop her. Billy was standing up this time and, being tall and strong, braced himself against Elsa's weight when she stood on her hind legs, resting her front paws on his shoulders, and nibbled at his ear. As soon as she released him I gave her such a beating that she sulkily left the tent and in a rather embarrassed way spent her affection now on Jespah, rolling with him in the grass, biting and clasping him exactly as she had done Billy. Finally, the whole family gambolled off towards the rocks. I do not know who was more shaken – poor Billy or myself. All we could think was that this extraordinary reaction of Elsa to Billy was her way of accepting him into the family, for only to her cubs and to us had she ever shown her affection in this way. Had she been jealous of Billy or disliked him she could easily have hurt him. We certainly did not want to risk a repetition of her demonstrations towards our friend, so we decided to break his visit short and leave camp immediately after breakfast. I was worried about George's malaria, but he assured us that he would be all right after a day or two, and, knowing from experience that these attacks never last long, we left with mixed feelings of regret and relief.

After a few miles we saw two elephants some thirty yards off the road. They tested our scent with raised trunks, made a few undecided, swaying motions and moved away. Ibrahim then walked along the track to see if all was safe for we were handicapped in our driving by the heavily loaded trailer which made any quick reversing in an emergency impossible. His reconnoitre saved us from driving straight into a single bull elephant who had remained on the road. We gave him time to move away, but he took much longer than we needed to take photographs before he disappeared into the bush. After that we continued without further excitement, if one discounts two punctures which landed us in a ditch. About two hours before reaching Isiolo the car stopped with an abrupt jerk. The trailer had lost one wheel and jammed its axle into the ground. There was nothing to do but leave our escorting Game Scout in charge of the wreck and send the lorry to tow it home. When we finally arrived at Isiolo it was well past midnight and by the time the house was opened up, our kit unloaded, bath water heated and supper prepared we were very tired and ready for bed. I felt very sorry for Billy, for his visit had been a constant strain and full of varying excitements.

In the meantime George, soon after we left, received another message asking again urgently for his presence in the same case. Malaria or not, he packed up camp next morning, but was interrupted by Elsa who suddenly appeared. She had kept away since her visits to Billy; now, feeling hungry, she came with the cubs. While they fed, George left for Isiolo.

8

The Camp is Burned

At the beginning of June, after ten days' absence, we returned to camp and, just before sunset, reached a place about six miles short of it. We saw that every tree and bush was loaded with birds of prey, and drove slowly towards them. Then suddenly we found ourselves surrounded by elephant who had closed in on us from every direction. It must have been the herd, numbering some thirty or forty head which had been in the neighbourhood for the past weeks. They had a large number of very young calves with them whose worried mothers came close to the car with raised trunks and fanning ears, shaking their heads angrily at us. It was a tricky situation and it was not improved by the arrival of my truck which, driven by Ibrahim, was following close behind us. George at once jumped on to the roof of the Landrover and stood there, rifle in hand. We waited for what seemed an endless time, then some of the elephants started to cross the car track about twenty yards from us.

It was a magnificent sight. The giants moved in single file, jerking their massive heads disapprovingly in our direction; to protect their young they kept them closely wedged between their bulky bodies.

After making infuriated protests, most of the herd moved away, leaving small groups still undecided in the bush. We waited for them to follow and eventually all but two went off; these stood their ground and seemed to have no intention of budging.

George wanted to see the kill which had attracted the birds and since the light was failing he decided to walk, with Makedde, between the two remaining groups of elephants. Meanwhile, Ibrahim and I stood on the roof of the car and kept a close watch on the beasts, so that we could warn George of their movements. He found a freshly killed water buck and lion spoor around it. Very little had been eaten, so plainly the lion had been interrupted by the arrival of the elephants.

When he returned the light was failing rapidly and the elephants still blocked our way. We could not drive round them, so we decided to make a dash for it and drove both cars past them successfully.

We wondered whether it might have been Elsa who had killed the water buck, but it was far from her usual hunting ground, and, besides, for her to tackle a beast with such formidable horns and heavier than herself (the buck must have weighed about four hundred pounds), while protecting her cubs, would have been a very dangerous enterprise, and we felt sure she would not have done such a thing unless she was very hungry indeed.

As soon as we got to camp we fired a signal to Elsa and then put up an aerial so that I could listen to my first broadcast about her which I had recently given at Nairobi. That night she failed to appear.

Next morning we started off very early to investigate the kill. Very little was left of it and the ground had been so trampled by elephant that we could not distinguish any spoor other than theirs. While we were creeping through some thorny bush we put up a rhino who crashed about, far too close for my liking. After a search which lasted for several hours we did find one pug mark of a lion cub. It could have been Jespah's but we did not believe he would have walked so far.

After our return to camp we were greatly relieved to see Elsa and her cubs on the Big Rock. As soon as she spotted us she rushed down and ended by throwing the whole of her weight against George who was squashed by her affection, then she bowled me over, while the puzzled cubs craned their heads above the high grass to see what was going on.

When we got back to camp we provided a meal for them over which they competed with such growls, snarls and spankings that we thought they must be very hungry. Little Elsa had the best of it and eventually went off with her loot, leaving her brothers still so hungry that we felt obliged to produce another carcase for them.

Later, while we were resting, Jespah, with surprising boldness, started chewing at my sandals and poking at my toes. As his claws and teeth were already well developed I quickly tucked my feet under me. He seemed most disappointed, so I stretched my hand slowly towards him in a friendly gesture. He watched it attentively, then looked at me and walked off.

That evening Elsa took up her usual position on the roof of the Landrover, but the cubs instead of romping about flung themselves on the ground and never stirred. As it was the hour at which they were usually most energetic, we were surprised. During the night I heard Elsa talking to them in a low moan and also heard suckling noises. They must indeed have been hungry to need to be suckled after consuming two goats in twenty-four hours.

In the morning they had gone. We followed their spoor and it led straight to the water-buck kill. So it must have been Elsa who two days earlier after a long stalk had tackled this formidable beast. It was hard luck on her that the arrival of the elephants had prevented her and the cubs from having a good meal out of her kill.

Now we understood why they had all been both so hungry and so exhausted when they came into camp.

We collected the fine horns of the water buck and hung them in the studio, a proud record of the cubs' first big hunt with their mother. They were now five and a half months old.

Our expeditions to visit Elsa on her Big Rock were always full of interest; at the far end, where the ridge was broken into deep crevasses and covered with clusters of candelabra euphorbias and low shrub, there were wonderful hideouts for all kinds of animals. The place literally swarmed with hyrax, which swished like shadows between the boulders. They would peep at us inquisitively and the two light markings above their eyes gave them such a questioning expression that I wondered whether their curiosity was ever satisfied. They blended completely with the colours of the rock and we had to use our field-glasses to follow their swift movements as they raced along its almost vertical face. Sometimes, growing bolder, a whole colony would gather together, having first posted a sentry who kept his eyes fixed on us. Secure under his guard they stretched out on the rocks and enjoyed the sun. Porcupine must have been about too, for we often found their quills.

The parrots were perhaps the most fascinating of all the inhabitants of that part of the bush.

One afternoon I watched a pair of them land on a baobab tree quite close to us, their short-tailed stumpy bodies displaying the most striking emerald and orange plumage as they hopped from bough to bough. Eventually they disappeared into a hole in a big branch; a few seconds later another parrot's head popped out of a hole close by. I saw that it belonged to an immature bird and hardly had it given a screech than the older pair emerged and settled close to the youngster. Very soon another young parrot came out of the hole; all four now perched close together chattering loudly.

Looking at the party through my field-glasses I noticed a third hole very near to the parrots' nests and within it a tiny face that looked almost human swaying from side to side, its enormous eyes and large ears showed that it belonged to a Bush-baby. These little beasts are nocturnal animals, and do not come out till after dark. They are so small that one can hold them in one's hands, but they have long bushy tails twice the length of their bodies.

One evening when Elsa and her cubs were walking back with us, she and Jespah got in front of us while Gopa and Little Elsa stayed behind. This worried Jespah very much; he rushed to and fro trying to marshal his pride, until his mother stood still, between us and him, and allowed us to pass her, thus reuniting the family. Afterwards she rubbed our knees affectionately as though to thank us for having taken the hint. That night a boiled guinea fowl disappeared from our kitchen. The cubs' father was the thief for we found his spoor by the kitchen tent.

The next morning I woke up to hear Elsa moaning to

the cubs in a nearby thicket. Since their birth we have never used the wireless when they were in camp so as not to frighten them. But to-day George turned on the morning news. Elsa appeared at once, looked at the instrument, roared at it at full strength and went on doing so until we turned it off. Then she went back to the cubs. After a while George tuned in again, whereupon Elsa rushed back and repeated her roars until he switched off.

I patted her and spoke reassuringly to her in a low voice, but she was not satisfied till she had made a thorough search inside the tent. Then she went to her family. I had often been asked how Elsa reacted to different sounds and had flattered myself that I knew how to answer these questions, but this reaction of hers was unexpected; before her release, when she was living with us, we had listened daily to the wireless, and though when we first tuned in she had always been startled, as indeed she usually was if I played the piano, as soon as she realized where the sounds came from she paid no attention to them. She differentiated between the engine of a car and of a plane. However loud the noise of the plane might be she ignored it, but the faintest vibration from a car engine alerted her, often before we heard it. I had tried singing to her to test her reactions, but whatever the melody I never observed any response. On the other hand, when occasionally I imitated the cubs' call in order to make her search for them she reacted at once as I intended she should, but if I did this for fun she paid no attention.

As a wild animal she could of course recognize various animal sounds and interpret the mood of the approaching beast. She could also sense our mood by the intonation of our voices. I think I am right in saying that she preferred a low voice in human beings to a high-pitched one, even where shrillness was not due to agitation.

On the night following the wireless incident I had plenty of opportunity for testing my own reactions to noise, as a herd of elephant disported itself between the river and our camp. The deep rumbling of their bellies, their trumpetings, the crash of falling trees and the splashing of water made sleep impossible. Then Elsa's lion added his roars to the concert. Improbably, I could also hear George snoring. Elsa took refuge in the thorn enclosure we had made round the tent which Billy had occupied.

In the morning the bush around the camp looked like a battlefield, the grass was trampled and there were deep holes made by the elephants' feet, but peace had returned and Elsa and her cubs lay in the dry lugga below the studio.

That afternoon we went back to Isiolo. We stayed there for nine days.

When on the 16th June we were making our way back to the camp, we nearly collided with two elephants which broke out of the bush just in front of our car. Luckily they seemed to be as frightened as we were and, when George jammed on the brakes, disappeared trumpeting into the darkness.

Elsa came into camp half an hour after we had fired a thunder flash. The cubs were with her. She gave us a great welcome, but I noticed that she had wounds on her head and chin and a deep gash on her right ankle which was very swollen. This must have been painful for she was not keen on moving more than was necessary and she refused to let me dress her cuts. The whole family were very hungry and it took two goat carcases to satisfy them.

Next morning we followed their spoor to see where they had laid up the night before we arrived. We knew it was on the far side of the river which she always preferred, though to us the two sides seemed identical. We were worried by her choice because we knew that the far bank was frequented by poachers and while, on her own, Elsa could not have been in any danger from them, with three cubs the situation was very different.

We had chosen the area in which we had released her because on either side of the river tsetse flies were very active in a belt a few miles wide. The bite of this species of tsetse is harmless to man and to most wild animals, but fatal to livestock, so we had good hopes that no tempting goats would come within Elsa's reach. She was very conservative in her habits, and though every two or three days she changed her lie-up she only moved around a very confined area, and this added to our reassurance.

Lately we had had plenty of evidence that neighbouring tribesmen were trespassing, so we felt it would be a good thing if we could identify the lie-up she was now most frequently using as this might enable us to come to her help if an emergency arose. We followed her spoor, which led us from the river, along a dry watercourse to a rocky outcrop about half a mile away from the camp, to what we called the Cave Rock. This contained a fine rainproof cavity with several 'platforms,' ideal resting places from which to survey the surrounding bush. Besides these amenities there were some suitable trees for the cubs to climb, growing nearby. This seemed to be Elsa's present lie-up.

When we got back to camp she and the cubs were waiting for us; she was nervous, but was very affectionate with me, allowing me to use her as a pillow; she also hugged me with her paws. Jespah, who had been watching us, apparently did not approve for after his mother had left he crouched and then started to charge me. He did this three times and though he swerved at the last moment, pretending to be more interested in elephant droppings, his flattened ears and angry snarls left me in no doubt about his jealousy. But it was significant that for his attack he chose a moment when his mother could not observe it. To placate him I gave him some titbits and then tied an inner tube to a ten-foot-long rope which I jerked about. While a tug-of-war was going on we suddenly heard the rumblings of elephant, which seemed to be having a game of their own in the studio.

The following day while we were having breakfast, four of these giants suddenly stood between the kitchen and the tents. They came so silently that we could have believed that they had fallen from the skies, and they left equally noiselessly.

Recently the crocodiles, who had scattered during the floods, had reassembled in the deep pools. This worried us because Elsa often took her meat down to the river before we had time to tie it up for the night, and several times after dark we had heard her growling, and, coming to her with torches and a rifle, found her defending her 'kill' against a 'croc' who invariably vanished as soon as we came on the scene. We tried to shoot some of them but we had only their eyes to aim at, as all the rest of their bodies was submerged, and as they have the most highly developed sense of impending danger of any wild animal I know, it is not surprising that, however carefully we stalked them, they always outwitted us.

On 20th June the cubs were six months old; to celebrate their first half year George shot a guinea fowl. Little Elsa, of course, took possession of it and disappeared into the bush. Her indignant brothers went after her but returned defeated and tumbling down a sandy bank landed on their mother. She was lying on her back her four paws straight up in the air. She caught the cubs and held their heads in her mouth. They struggled to free themselves and then pinched Mum's tail. After a splendid game together, Elsa got up and walked up to me in a dignified manner and embraced me gently as though to show that I was not to be left out in the cold. Jespah looked bewildered. What could he make of this? Here was his mother making such a fuss of me, so I couldn't be bad, but all the same I was different from them. Whenever I turned my back on him, he stalked me, but each time I turned and faced him he stopped and rolled his head from side to side, as though he did not know what to do next. Then he seemed to find the solution; he would go off; he walked straight into the river evidently intending to cross to the other bank. Elsa rushed after him. I shouted, 'No, no,' but without effect and the rest of the family quickly followed him. Young as he was Jespah had now taken on the leadership of the pride and was accepted by the family.

When they returned Elsa dozed off with her head on my lap. This was too much for Jespah. He crept up and began to scratch my shins with his sharp claws. I could not move my legs because of the weight of Elsa's head resting on them, so in an effort to stop him I stretched my hand slowly towards him. In a flash he bit it and made a wound at the base of my forefinger. It was lucky that I always carry sulphanilamide powder with me so I was able

111

to disinfect it at once. All this happened within a few inches of Elsa's face but she diplomatically ignored the incident and closed her eyes sleepily.

I stayed on watching the last glow of the sinking sun gild the tips of the doam palms, then all colour vanished and darkness fell.

Suddenly I saw Little Elsa stiffen and stare intently at the opposite bank of the river. I looked and there was a large bull elephant, going down to the water to drink. I poked Elsa, who now noticed the elephant but made no move. I watched him raise his trunk and scent. He remained in this position for a long time, but as we were down wind he gained no knowledge of our presence. In case he should decide to cross the river I freed myself from Elsa's weight and brought the cameras into safety, but by the time I had taken them to the tent, the elephant had begun to move and was slowly disappearing into the bush.

After this we all returned to camp and Jespah seemed so friendly that I began to wonder whether when he bit me it was only in play. Certainly, between himself and his mother, biting was a proof of affection.

By now we were, however, beginning to worry about his relationship to us. We had done our best to respect the cubs' natural instincts and not to do anything to prevent them from being wild lions, but inevitably this had resulted in our having no control over them. Little Elsa and her timid brother were as shy as ever and never provoked a situation which required chastisement. But Jespah had a very different character, and I could not push his sharp, scratching claws back by saying, 'No, no,' as I used to do when Elsa was a cub and so taught her to retract her claws when playing with us. On the other hand, I did not want to use a stick. Elsa might resent it if I did and indeed she might cease to trust me. Our only hope seemed to lie in establishing a friendly relationship with Jespah, but for the moment his variable reactions made a truce more possible than a friendship.

After five days in camp we returned to Isiolo and, when we reached home, found that in a short time it was going to be necessary for George to go to the north for a three-week safari. We did not wish to desert Elsa for so long, and as in the absence of George and his Landrover, I should not be left with enough transport to go backwards and forwards between Isiolo and the camp, I decided that I would spend these three weeks in the bush, even if it upset the cubs' wild life.

Before setting off I had two weeks by myself at Isiolo after which I planned to meet George in the first week of July at the camp. He would then be returning from patrol and on his way to Isiolo to get ready for the safari to the north.

As I approached the camp I was worried because I did not see George and drove on filled with foreboding which was increased when, as I drew nearer, the air became so full of smoke that my lungs were stinging.

When we arrived I could hardly believe my eyes. The thorn bushes were in ashes and smouldering tree trunks added to the grilling heat. The two acacia trees which provided shade and were the home of many birds were scorched. In the charred and blackened scene the green canvas of the tents stood out in sharp contrast. I was much relieved when I found George inside one of them eating his lunch.

He had plenty to tell me. When he had arrived, two

days earlier, he had found the camp burning and seen the footprints of twelve poachers. Not only had they set fire to the trees and the thorn enclosure but they had also destroyed everything they could find. They had even uprooted the little vegetable garden that Ibrahim had planted.

George had been very worried about Elsa and had fired several thunder flashes between seven and ten p.m. without getting any response. Then at eleven she and the cubs had suddenly appeared, all ravenously hungry. Within two hours they had eaten an entire goat. Elsa had been most affectionate and had several times come to lie on George's bed during the night: he noticed that she had several wounds. She left at dawn; soon afterwards he followed her spoor and eventually saw her sitting on the Whuffing Rock.

Then he went off to try and discover where she had come from on the previous evening. Her spoor which led down from the river was mixed up with the footprints of the poachers. He wondered whether they had been hunting Elsa and the cubs.

After lunch he sent three Game Scouts to search for the camp burners. They returned with six of the culprits. He kept them busy rebuilding the camp, which was no agreeable task, considering the amount of thorny bush which they were obliged to cut for our enclosures.

Elsa and her cubs who had spent the night in camp left soon after daybreak. Half an hour later George heard roars coming from the direction of the Big Rock, which was the way they had gone, so he assumed it must be Elsa; he was therefore much astonished to hear her voice coming from across the river soon afterwards. Then she appeared wet and without her cubs and seemed very agitated: she had several bleeding marks on her hind-quarters.

In a few minutes she left hurriedly, rushing towards the Big Rock calling loudly. George felt sure that she must recently have had an encounter with an enemy for her wounds were not made by a quarry; also, her nervous state suggested that she knew that whatever beast had threatened her was still in the neighbourhood. George now thought that the roars he had first taken for Elsa's were probably those of some fierce lion who had attacked her and that while the two were fighting the cubs had scattered and after the battle Elsa had escaped across the river. Now he followed Elsa in search of her family.

Together they climbed up the Big Rock. When they got to its top Elsa called in a very worried tone of voice. Of the cubs there was no sign. George, and Elsa, searched back and forth between the rocks and the camp. Suddenly she became much interested in a patch of dense bush which she sniffed attentively and then called towards. George investigated it; he saw no sign of the cubs inside the thicket, but Elsa remained beside it while he went back to camp to collect Nuru to join in the search. All morning they looked for spoor but found only Elsa's pug marks, which must have been made in the early morning. These showed that she had gone quickly towards the river and crossed it below the studio.

After a long fruitless trek George sent Nuru back to camp and carried on alone until he found Elsa at the base of the Whuffing Rock, still calling desperately for her children. Together they crept along the ridge, looking into all possible hideouts. They found the spoor of a large lion and of a lioness and Elsa seemed most upset. During the morning she had insisted on taking the lead, but now she was content to follow George.

When they reached the end of the rock, near to the place where the cubs were born, Elsa sniffed very persistently into a cleft. Suddenly George saw one cub peeping over the top of the rock above them and soon another appeared; they were Little Elsa and Gopa. Jespah was missing.

When they saw their mother they rushed down and rubbed noses with her and finally went off with her towards the 'kitchen lugga.' All this had taken place just before I had arrived and as soon as he had finished his lunch George intended to look for Jespah. Naturally I went with him.

After about an hour Elsa appeared at the foot of the Big Rock and gave me a most heartening welcome. As I was brushing off the tsetse flies from her coat and dressing her wounds, the two little cubs peeped at me from a distance of about sixty yards and then ran off. When I began to rub the M. and B. powder into Elsa's injuries I found that not only had she gashes on her hind-quarters but very nasty tears on her chest and chin.

While all this was going on the cubs remained in the bush and Elsa paid no attention to them. To encourage them to come to their mother we retired behind some rocks and after a while they rushed to her.

As soon as they were safely settled on the top of the ridge, George went off to search for Jespah by the Zom rocks, while I investigated the foot of the range. Looking back at Elsa I noticed that she was pulling a grimace and scenting in the direction of the thicket which George said had interested her so much in the morning, but when I called to her she did not budge. The ground was covered with fresh lion spoor, so I understood why she was frightened. However, after George returned, she and the two cubs joined us below the rock.

Now she trotted ahead of us towards the interesting thicket. Just after she had passed it I suddenly saw that not two, but three cubs were scampering behind Elsa, in the most casual manner. Jespah's reappearance after a day's absence seemed to be taken by the family as the most natural thing in the world. We, however, were greatly relieved and followed them to the river where they stopped for a long drink, while we went ahead to prepare a carcase for them in camp. When finally we were able to sit down and enjoy our dinner we discussed Elsa's curious behaviour. Why had she not persevered in the search for Jespah? Had she known all that time that he was hiding in the thicket? But was this likely? Why should he have remained alone for twelve hours only a very short distance from the camp, the river and the rocks where the rest of his family were; and why had he not answered his mother's call and ours?

Had the strange lions still been near the rocks, this would have explained Elsa's fear and Jespah's, but had this been the case it was unlikely that the other two cubs would have chosen to take refuge there.

After dinner George had to start back for Isiolo to prepare for his three weeks' safari. I was not very happy to see him go at this late hour, when all the wild animals were on the move.

Soon after he had left the lions began to roar from the Big Rock and kept on calling for most of the night. Elsa when she heard them at once moved herself and the cubs as near as possible to my enclosure and stayed there till dawn; then she took them across the river. Later I saw their pug marks on the sandbank just below the camp. They were mixed up with buffalo spoor; this beast remained near us for some time, did not seem to be troubled by our presence and came each night, just below the tents, to drink.

During these days I made some attempts to shoot crocodile but without much success.

Elsa and her cubs were well aware that the 'crocs' were not friendly and often watched the water attentively for any suspicious eddy or floating sticks. On the other hand, their reactions were inconsistent, and I was anxious about their safety.

One afternoon I called to Elsa, who was on the far bank. She appeared at once and was preparing to swim across with the cubs, when suddenly they all froze and stared intently into the water. Then Elsa took the cubs higher up the river and they appeared opposite the 'kitchen lugga.' Here the water is very shallow in the dry season. In spite of this they did not cross for an hour, nor did the cubs indulge in their usual splashing and ducking games. This was reassuring for it showed their prudence, but it was characteristic of their variable reactions that next day when I called Elsa from the same place at the same time, they all swam across at once, and without the slightest hesitation. Then I noticed that Elsa had a wound the size of a shilling in her tongue, and a very deep gash across the centre which was bleeding. This did not prevent her from licking the cubs, which surprised me.

When it was getting dark we were all sitting near to the river. Suddenly Elsa and her cubs looked at the water, stiffened and pulled grimaces and three or four yards away

I saw a 'croc.' I knew that he must have been a big fellow for his head was about a foot long.

I fetched my rifle and killed him. Although the cubs were less than three feet from me, the shot did not upset them. Elsa afterwards came and rubbed her head against my knee as though to thank me.

Nearly every afternoon she brought her cubs to the sandbank. Among its attractions were fresh buffalo droppings and sometimes elephant balls as well; in these they rolled to their great satisfaction. The cubs also played on the fallen palm logs. There was no question, when they fell off as they frequently did, of their landing on their feet, like the proverbial cat; on the contrary they fell clumsily on to the grass like a dropped parcel and seemed most surprised at their abrupt descent.

It was about this time that Jespah became more friendly. Now he sometimes licked me and once even stood on his hind legs to embrace me. Elsa took great care not to show too much affection for me in the presence of the cubs, but when we were alone was as devoted as usual. Her trust in me was as complete as ever and she even allowed me to take her meat from her claws and move it to a more suitable spot when I thought this necessary. She also permitted me to handle the cubs' meat. For instance, in the evening when I wanted to remove a partly eaten carcase from the riverbank so that the 'crocs' should not finish it off, she never interfered, even if I was obliged to drag it over her, and, still more remarkable, even when the cubs were hanging on to it and defending it.

At dusk the cubs were always full of energy and played tricks on their mother which made it hard for her to retain her dignity. Jespah, for instance, discovered that when he stood on his hind legs and clasped her tail she could not easily free herself. In this fashion they would walk round in circles, Jespah behaving like a clown until Elsa had had enough of it and sat down on top of him. He seemed to be delighted by her way of putting an end to the game and would lick and hug his mother until she escaped into our tent.

But it was not long before the tent ceased to provide her with an asylum, for he followed her into it, giving a quick look round and then sweeping everything he could reach to the ground. During the night I often heard him busily engaging in sorting through the food boxes and the beer crate; the clattering bottles provided him with endless entertainment. One morning the boys found fragments of my precious rubber cushion in the river; but I really could not blame Jespah for this as I had stupidly forgotten to remove it from my chair the evening before. He became quite at home in the tent, but his brother and sister were less venturesome. They stayed outside watching the fun.

One evening Jespah even visited the boys in the kitchen. He appeared while they were sitting around an open fire, walked round them sniffing, inspected the place thoroughly and then went off.

As a result of the burning of the camp there were no scorpions to plague us this season. As a rule, we were obliged to keep a stick handy, so that when, after dark, they scuttled across the tent floor with raised tails, we could kill them. It always surprised me that, so far as we could tell, neither Elsa nor the cubs had ever been stung by them. Some years earlier my Cairn terrier, who was about the size the cubs had now reached, nearly died as a result of a sting; and I myself had once been stung by a scorpion not more than an inch long and had afterwards suffered from the most painful convulsions and swollen glands, until the poison, for which no antidote had been discovered, got absorbed. There are two kinds of scorpions in our area, one which is black and about four inches long; it looks most forbidding, but, in fact, the smaller pale-coloured species has a much more effective poison.

Unfortunatley, the fire which had rid us of the scorpions also disposed of the friendly frogs, who used to come every night to catch insects by our lamp-light. One particu-larly bold one used to hop about on top of Elsa when she was resting in the tent. It always surprised me that they took no notice of each other. When I had my canvas bath tub filled jolly frogs used to jump around it at once, and I had got so used to them that now I missed them.

9

Elsa's Fight

One morning Makedde observed vultures circling and, going to the spot about a mile downstream, found the remains of a rhino which had been killed by poisoned arrows the day before while drinking.

The poachers had left plenty of footprints and had erected machans on trees close to the drinking place. They must have been well informed and known that I was alone with only Makedde to guard the place for, had George been with me, they would never have dared to indulge in these activities so close to the camp.

On the night of the 8th July there was quite a concert, Elsa's lion 'whuffing,' a leopard coughing and hyenas howling. The next evening while I was taking tsetse flies off Elsa as she sat in my tent with her head on my lap, I was startled by a great roar from her lion. Like a flash she dashed off in the direction of the 'kitchen lugga.' The cubs rushed after her but soon returned and sat looking bewildered outside the tent. Later, Elsa came back and stayed in camp until the lion ceased to call. As soon as she had gone, I heard the crackling of bones and realized that the hyenas were feeding. Elsa did not come back for more than twenty-four hours, although during the following night her mate shook the air with his roars. To my surprise she and the cubs appeared at about nine o'clock next morning. They asked for food and when they got a carcase Elsa dragged it to the river; she left two hours later after eating most of it. Did her arrival at this unusual hour mean that she was avoiding her mate and only came to feed when she felt that he would not appear?

That evening she brought the cubs with her. After I had gone to bed she set off three times to cross the river, but as I did not see why I should be obliged to provide free meals for any predator who happened to be in the neighbourhood, I called her back each time and insisted that she should guard the remains. She obeyed and only made a final departure just before daybreak, when the carcase needed no more protection.

For three days she arrived in camp long after dark, and on the fourth (the 15th of July) brought only two cubs; Jespah was missing. I was very worried so after waiting for some time, I began repeating his name over and over again, till Elsa decided to go upstream and look for him, taking the two cubs with her.

For over an hour I heard her calling, till the sound gradually receded into the distance.

Then suddenly there were savage lion growls, accompanied by terrified shrieks of baboons. As it was dark I could not go to see what was happening and awaited the outcome feeling miserable, for I was sure that Elsa was being attacked by lions.

She came back after a while, her head and shoulders covered with bleeding scratches and the root of her right ear bitten through. There was a gap in the flesh into which one could stick two fingers. This was much the worst injury she had ever suffered. Little Elsa and Gopa came back with her and sat a short distance away looking very frightened. I tried to put sulphanilamide into Elsa's wounds but she was far too irritable to let me come near her, nor was she interested in the meat which I brought her. I placed the carcase half-way between myself and the cubs. They pounced on it, dragged it into the dark and I soon heard them tearing at it.

I sat a long time with Elsa; she held her head on one side and the blood dripped from her wound. Eventually she rose, called the cubs and waded across the river.

I could hardly wait till it was light to go and look for Jespah. Next morning, following Elsa's spoor, Makedde, Nuru and I went to the Cave Rock and were much relieved to find the family reunited. I was happy to know that Jespah was safe and that I could now concentrate on treating his mother. The wound in her ear was still bleeding profusely, and at intervals she shook her head to drain the cavity. Owing to its position she could not lick the wound, but scratched constantly to keep off the flies; none of this was likely to improve the cleanliness of the wounds.

All the cubs seemed very subdued though Jespah licked his mother affectionately.

The boys stayed out of sight while I tried to put M. and B. into the injury, but Elsa was not co-operative and each time I approached her head she moved away, apparently with considerable effort. Suddenly I was startled to hear voices. I thought they were probably those of poachers. I had to think quickly. Was it best to stay put? Probably not, for Elsa did not seem to want our company and might well go off with the cubs and fall into the poachers' hands. I went back to camp, hoping that as she must be hungry she would follow.

We made a detour on our return journey, so as to inspect

A lesson in combat

Elsa suckling cubs at nine months

Unlike Elsa, Jespah could not always be relied on to retract his claws!

Jespah (left), Gopa (right), Little Elsa (behind)

Little Elsa, the wildest of the three cubs,
always kept her distance and often climbed trees

An evening drink, October

I came upon Elsa trotting along with the cubs

As the cubs would not cross the swollen river
for fear of the crocodiles, Elsa swam across
to them with a carcass

Family portrait

Elsa, Christmas 1960

the previous night's battlefield. We found it on a sandbank in the middle of the river, about half a mile from the camp. There were plenty of lion pug marks mixed up with baboon spoor, but though we could distinguish the imprints of one male lion we could not be sure whether he had been alone or not.

I waited anxiously till the late afternoon for Elsa and her family to arrive. I then managed to introduce some M. and B. tablets into the meat which she took from my hand. I thought that if I could get fifteen tablets down her daily, there was a good chance that her wound would not go septic. Her ear drooped, suggesting that the muscles had been injured and she constantly shook her head to get rid of the oozing liquid.

Jespah, who had been the cause of the encounter, was very friendly. He licked me and several times tilted his head looking straight at me for a long time.

There is a belief that the members of the cat tribe can never look one in the face for any length of time; this is not true of Elsa and her sisters or of her cubs. Indeed, I found that they convey their feelings by the varying expressions of their eyes, far more explicitly than we do in words.

After Elsa had settled down for the night a lion began calling. This seemed to alarm her and she shortly afterwards went off with the cubs.

I was glad when they all returned during the following afternoon; Jespah occasionally poked his nose into my back, in a friendly fashion, but apparently Elsa did not approve for she placed herself between him and me.

Towards evening Nuru herded the goats towards the truck. This was the first time I saw the cubs take any interest in them. We had, of course, been careful to avoid any contact between the cubs and living goats and they had never before reacted to their bleating.

I woke up during the night and smelt smoke, and, getting up, saw that the river-bank below the kitchen was on fire. I called the boys, and as the fire had begun so close to the water, they had it out in no time. It had been started by a smouldering tree stump which had been overlooked at the time of the original fire and had ignited the dry bush around it.

Elsa watched all the fire-fighting activities from the tent and kept on calming the cubs with her reassuring '*mhn-mhns*.' When all was quiet again they crossed the river.

Soon afterwards I heard two lions grunting as they cracked the bones of the carcase which was lying in front of George's tent. They spent a long time over their meal and only went off at dawn when the boys began talking in the kitchen. Then they crossed the river accompanied by the barking of baboons, to which they replied by loud 'whuffings.' We found the spoor of a large lion and of a lioness.

Elsa kept away for some days. I thought her absence was explained by the presence of this pair who had remained nearby, and who the following night grunted round the goat truck.

The boys and I made several searches for Elsa, but these were unsuccessful; while so occupied, we put up a rhino and a few buffalo.

After Elsa had been absent for four days I became very anxious, for her wound must be a very big handicap to her in hunting, and I was afraid also that the poachers might do her some harm. When on the evening of the 20th

July I saw vultures circling, my heart sank. We went to investigate but all we found was more evidence of the poachers. They had made hides near to every drinking place, on both sides of the river. We also found the ashes of recent fires and charred animal bones.

A week earlier when Makedde had found poisoned arrow-heads in a rhino, I had sent a message to the Warden of the reserve, asking him to send Scouts to patrol the area. Now on our return to camp we found that they had arrived and I was very glad to see them. With our reinforcements we set out next morning to look for Elsa and arranged that if anyone spotted her they should fire a shot.

Three hours later I heard a report and returned to camp to be told by two of the newcomers that they had seen Elsa and the cubs under a bush on the opposite side of the river, about a mile inland.

She was lying in the shade and the cubs were asleep. She had seen the men approach but had not moved. This sounded odd, unless she was so ill that she did not care if even strangers were close by.

Makedde suggested that we should take some meat to her, but not enough to satisfy her hunger, and so tempt her to come back to camp. As we approached her lie-up I signalled to the men to stay behind and called to her.

She emerged, walking slowly, her head bent low to one side. I was surprised and alarmed that she should have settled in such an exposed place where she could easily be seen by poachers. I noticed that her ear had gone septic and was discharging pus; she was obviously in great pain and when she shook her head, as she did very often, it sounded as if her ears were full of liquid. Besides this, both she and Little Elsa were covered with blowflies. I was able to rid Elsa of hers, but the cub was far too wild to let me help her. Meanwhile, she and her brothers fought over the section of carcase we had brought them and soon there was nothing left for Elsa but polished bones. She looked on resignedly and certainly gave the lie to the well-established legend that lionesses gorge themselves and let their cubs go hungry. Jespah thanked me for his meal by licking my hand with his rough tongue. I tried to induce Elsa to come back to camp by calling, 'maji, chakula, nyama,' but as she did not move, went home without her.

As I had taken a lot of photographs I went to the camp to get another film; then I heard the cubs arrive on the opposite bank and took a short-cut down to the river. Suddenly Elsa broke out of a bush and knocked me over. She obviously was suspicious that I had returned from a different direction, and feared for her cubs. She had been nervous all the afternoon and was plainly in pain for whenever the cubs accidentally touched her ear she snarled and cuffed them irritably. Jespah seemed aware of her state and constantly licked her.

That night after I had gone to bed – Elsa and the cubs had left the vicinity soon after – I heard a leopard cough and a lion roar. I got up and called to the boys to open my thorn enclosure so that I could go out and put the remains of the meat into my car. I did not wish to encourage all the predators in the neighbourhood to share Elsa's food supply and in doing so drive her away.

So as soon as her ear had healed and she could hunt, I was determined to leave her. By now I had been three weeks alone in camp and George was overdue. I wished he would return soon for when his tent was occupied the predators never came near the meat which was tied up close to it. In his absence wild lions prowled round the camp every night and although Makedde and Ibrahim could have used their rifles if an emergency arose, I was nervous about the safety of the boys.

At last George arrived and was greeted by the roars of a strange lion. Hearing that Elsa had not been seen for several days, he decided to go and look for her, and he was also determined to try to scare off the strange lion, and his fierce lioness who had so often injured Elsa. We knew her and her mate quite well by now; at least by voice, and we were also familiar with their spoor. They ranged along the river for about ten miles. Of course they shared the country with other lions besides Elsa, but she was the only one who kept permanently to the vicinity of the camp. The fierce lioness had lived in this region long before Elsa but we did not know what she had done to displease this disagreeable beast. We were pretty sure that she had not competed for the attention of her mate, but had kept strictly to her own young lion. Perhaps Elsa had interfered with her hunting or her territorial claims, or perhaps the creature was just bad-tempered. Anyway, we were sure now that she had chased Elsa and the cubs over the river and towards the poachers and that she and her mate had, for several days, taken over the Big Rock.

Tracking on the far side of the river we eventually found the cubs' pug marks leading into a large group of rocks which we called the Border Rocks, as they were at the boundary of Elsa's territory, but by then it was too dark to do anything except go home. When we returned next morning we found the fresh spoor of a lion and a lioness superimposed upon the cubs' imprints. We were full of hope until we saw that the spoor led so far away that it was unlikely to have been made by Elsa. On our way home we observed a drop-spear trap close to the river. It was suspended from a tree which overhung the game path.

The drop-spear trap is a deadly device consisting of a log about one foot in diameter and two feet in length; to the cross section which faces the ground is attached a poisoned harpoon. When the log is released it falls upon the animal passing below and its weight is sufficient to ensure that the harpoon penetrates the thickest hide.

By the time we reached the camp we had identified the spoor of five different wild lions and all night we were kept awake by their roarings.

Next day we searched upstream on the far side of the river. Here, too, there were plenty of lion pug marks – including those of a lioness with three cubs. They led us five miles from camp to a part of the bush which, so far as we knew, Elsa had never visited. As we approached a baobab tree, we heard the sound of startled animals bolting and the Toto caught a glimpse of the hindquarters of a lion and of three cubs which could have been Elsa's. They were gone in a flash and though we called and called there was no response.

George and I followed their tracks for some way, but we were puzzled; if they were Elsa's family why had they rushed away from us? On the other hand, was it likely that there was another lioness about with her three cubs of around the same size as Elsa's? On our way back we found fresh spoor of a lion leading in the direction we had just come from.

Next morning we returned to this place and within five hundred yards saw some very recent spoor of a lion, a lioness and cubs. This led us up a dry watercourse, then towards some rocks, but before reaching them, the pride had abruptly turned back, run fast to the river and crossed it.

The pug marks on the far bank were still wet. It was plain that having heard us the pride had bolted. All we could be sure of was that they had scattered and run very fast.

After two hours' tracking we found that the pride had reassembled in a sandy watercourse. We kept very quiet till we heard the agitated barking of baboons and simultaneously the roar of a lion. He was very close to us.

His voice was familiar to us for we had often heard it at night. He sounded hoarse and the boys used to say that he must have malaria.

George proceeded to stalk him and we came so close that I was nearly deafened by his next roar. Suddenly I caught sight of his hindquarters only thirty yards away and the boys actually saw his head and mane.

It is most unusual for a lion to roar at eleven in the morning. This one was evidently calling to a lioness, whom presently we heard replying from the direction of the barking baboons. Hoping it might be Elsa, we by-passed the hoarse lion and had a good look round, but saw nothing.

Finally, tired and thirsty, we sat down and made tea at the place which appears as the frontispiece in *Born Free*. Here we discussed the two possible explanations of Elsa's disappearance. Rather than stay in camp and risk being mauled by the ill-tempered lioness, she might have decided to share the hazards of the hoarse lion's life, whose spoor might have been the one we found the previous day. That was an optimistic solution to the mystery; a pessimistic alternative was that Elsa had died of her septic ear and that the cubs had been adopted by a pair of wild lions.

On our way back we saw flocks of vultures around the 'kitchen lugga' and the boys went ahead to inspect the kill.

I hung back dreading to learn what they had discovered, but soon they shouted that they had found the carcase of a lesser kudu, which had probably been killed during the night by wild dogs.

We spent the next two days covering the boundaries of Elsa's territory, partly on foot and partly by car. We searched in particular for drinking spoor.

Eventually we found some cub spoor downstream, but if this was the spoor of the cubs who had drunk upstream, then they must have covered at least fifteen miles in two days – probably much more as they had not kept to the river-bank.

We searched on an average for eight hours a day. We learned nothing of Elsa, but a lot about the poachers. We destroyed many of their hides and in one found a bit of rope which I had used to fasten the wicker gate of my tent enclosure. Indeed, we saw so much evidence of their activities that George decided to send immediately for an anti-poaching squad and determined that as soon as he could he would establish a permanent Game Scout Post on the river.

George left in the last week of July and I continued to search for Elsa, and the next morning, walking with Makedde along the car track towards the Big Rock, traced the spoor of a single lion who had evidently come towards the camp; I saw also the imprints of pointed shoes, which Makedde recognized as identical with those which he had seen near the piece of rope in the poachers' hideout. Both spoors were superimposed on the tyre marks of George's car.

Plainly the poachers were keeping an eye on our movements, and no doubt, having heard George's car go off, had next morning come to reconnoitre. How disappointed they must have been to discover that I was still in residence.

It was very hot and, after several hours of tracking, Makedde and I sat down to rest.

My spirits were very low. It was now over a fortnight since the fierce lioness had attacked Elsa and except for the occasion when the Game Scout had found her in the bush, she had not been seen, nor had there been any trace of the cubs. I was particularly worried because, during the time in which I had observed Elsa's wounds, instead of healing their condition had grown worse. In such a state, could she, I wondered, hunt and provide food for herself and for the cubs? Also, the presence of the poachers provided another and perhaps even more serious cause for anxiety.

Feeling miserable, I asked Makedde whether he loved Elsa. He looked startled but replied warmly: 'Where is she that I could love her?' This made me even more depressed. Makedde, watching me, scolded more angrily: 'You have nothing but death in your mind, you think of death, you speak of death and you behave as though there were no Mungo (God) who looks after everything. Can't you trust him to look after Elsa?'

Encouraged, I got up and went on with the search; but two days passed without bringing any result.

On the evening of the sixteenth day since Elsa and the cubs had disappeared, after lighting the lamps I poured myself a drink and sat in the dark straining my ears for any hopeful sound. Then, suddenly, there was a swift movement, and I was nearly knocked off my chair by Elsa's affectionate greeting. She looked thin but fit and the wound in her ear was healing from the outside, though the centre was still septic. Plainly she was hungry for when the boys came towards us with the carcase I had asked for, she rushed at them. I yelled, 'No, Elsa, no.' She stopped, obediently returned to me and controlled herself until the meat had been attached to a chain in front of the tent, then she pounced on it and ate voraciously. She seemed to be in a great hurry, gorged herself on half the goat and then withdrew out of the lamplight and cunningly moved farther away till she finally disappeared in the direction of the studio.

I was immensely relieved to know that she was well, but where were the cubs? Her visit had only lasted half an hour and I waited long into the night hoping that she might return with them to finish off the goat. As this did not happen, I eventually carried the remains into my car to save them from being eaten by predators, and went to bed.

At dawn on the 1st August I was woken by the miaowing of the cubs and saw them crawling close to my thorn enclosure. I called to the boys to bring the meat and joined Elsa who was watching her youngsters fighting over the meat.

It was soon obvious that what remained of Elsa's last night's supper was not going to satisfy four hungry lions, so I ordered Makedde to kill another goat and managed to keep Elsa quiet while this was going on. Her self-control was astonishing, and only when the men dropped the carcase within ten yards of her did she get up and drag it into the bush near the river.

Little Elsa and Gopa followed her, but Jespah was far too busy crunching bones to pay any heed to what was going on and only after he had been on his own for some time did he decide to join the family, and straddling what was left of the old 'kill' he took it down to the river.

I sat under a gardenia bush close by waiting my chance to introduce some medicine into Elsa's meat, to help her septic ear heal. I was relieved, but puzzled, not to see a single new scratch on her or the cubs, though they must have hunted during all these days when they were absent from camp.

The cubs growled, snarled and cuffed at each other for the best bits of meat. Living in the bush had certainly made them become more wild, for now they were constantly on the alert for suspicious sounds and nearly panicked when some baboons barked.

The two little cubs were shyer than ever and were frightened if I made the least movement, but, to my surprise, Jespah came up to me, tilted his head on one side with a questioning look, licked my arm and plainly wished to remain friends.

The sun was high, it was getting hot, and so when the cubs had eaten all they could they had a splendid game in the shallows, ducking, wrestling, splashing and churning up the water till at last they collapsed in the shade on a rock, where Elsa joined them.

As I watched them dozing contentedly with their paws dangling over the boulder I humbly remembered Makedde's reprimand for my lack of faith – a happier family one could not wish to see.

In order to try to discover what they had been up to during their long absence I had asked him to follow the spoor which Elsa had made when she had arrived in camp.

Meanwhile, I dressed her wound while she was too sleepy to object to the treatment. When it got dark I went to the tents to hear Makedde's report.

He told me he had traced her to the limit of her 'territory' and that there, on some rocky outcrops, he had found not only her pug marks and those of the cubs, but also the spoor of at least one lion, if not two.

This probably explained how she and the cubs had been fed and also accounted for her strange behaviour when she was surprised by the Game Scout and us, for her reactions were typical of a lioness in season.

It may seem odd that this solution had not occurred to us but as Elsa was still suckling her cubs we had not expected her to be interested in a mate. We had accepted the general belief that wild lionesses only produce cubs every third year, because in the interval they are teaching the young of the last litter to hunt and become independent. Could Elsa have returned more quickly than we expected to breeding condition because of the food we had supplied? Certainly at seven and a half months they could have survived on a meat diet and obviously she could not know that we were only staying on so as to treat her wounds and help her to get fit and able to teach her cubs hunting.

10
Dangers of the Bush

At about nine that evening Elsa and the cubs came from the river, settled themselves in front of my tent and demanded their supper. As the remains of the meat was still by the gardenia bush I called to Makedde and the Toto and asked them to come and help me drag it in. I collected a pressure lamp and we went down the narrow path which we had cut through the dense bush from the camp to the river.

Makedde, armed with a stick and a hurricane-lamp, went ahead, the Toto followed close behind and carrying my bright light I brought up the rear. Silently we walked a few yards down the path. Then there was a terrific crash, out went Makedde's lamp and a second later mine was smashed as a monstrous black mass hit me and knocked me over.

The next thing I knew was that Elsa was licking me. As soon as I could collect myself I sat up and called to the boys. A feeble groan came from the Toto who was lying close to me holding his head, then he got up shakily, stammering, 'Buffalo, buffalo.' At this moment we heard Makedde's voice coming from the direction of the kitchen; he was yelling that he was all right. As we pulled ourselves together the Toto told me that he had seen Makedde suddenly jump to the side of the path and hit out with his stick at a buffalo. The next moment the Toto had been knocked over and then I had been overrun. What had happened when Elsa and the buffalo met face to face none of us will ever know. Luckily the Toto had no worse injury than a bump on his head, caused by falling against a palm trunk. I felt blood running down my arms and thighs and was in some pain, but I wanted to get home before examining my wounds. This incident certainly belied the popular belief that a lion however tame becomes savage at the scent or taste of blood.

Elsa, who had obviously come to protect us from the buffalo, seemed to realize that we were hurt and was most gentle and affectionate.

I had no doubt as to the identity of the buffalo, since for several weeks past we had seen the spoor of a bull buffalo, going from the studio through the river bush to the sandbank, where a triangular line of impressions marked his drinking place. After quenching his thirst he usually continued upstream, passed below the kitchen and then settled for the day on a thickly wooded island about half a mile away.

During our walks we had often 'put him up' but, though our camp was within his territory, until now we had never had any disagreeable encounter with him; he had never come out for his drink till well after midnight, and it was only in the early hours of the morning that we had heard his snortings and splashings.

This evening he must have been unusually thirsty and come out very early. Probably Elsa had heard him on the move and that was why she had brought the cubs into the camp at nine. When he saw us come down to the river with our lamps the buffalo had evidently been frightened and rushed up the nearest path to safety, only to find us blocking his way.

I received several kicks which left their marks on my thighs and I could only feel very thankful that they had not landed on more vulnerable parts of my anatomy.

Elsa came back with us to camp where we found the cubs waiting for her; how she had prevented them from following her puzzled me.

I was worried about Makedde and went at once to the kitchen to see what condition he was in. There I found him, unhurt and having a splendid time, recounting to his awestruck friends his single-handed combat with the buffalo. I am afraid his heroic stature was slightly diminished by the appearance of my bleeding legs, but the main thing was that we were all safe.

I spent a very uncomfortable night, for as well as my painful wounds all my glands began to swell and it was difficult to find a position in which I could relax, or to breathe without increasing the discomfort of my aching ribs. All the same, I felt rather elated at having acquired a very perfect buffalo autograph in the shape of the exact imprint of his hoof. I also had a curious feeling that the encounter had some significance. Earlier that evening while watching Elsa and her cubs playing happily after their sixteen days' absence, I had said to myself that I was completely happy, and now the day had been sealed by a warning not to tempt Providence.

In the morning my pelvis, thighs and arms displayed all the colours of the spectrum. It took three days for the pain to stop and for the swollen glands to go down, and it took a much longer time for the hoof marks and the sepsis at the centre of them to disappear.

The next afternoon Elsa took great care to drag her 'kill' a long way upstream and to straddle it across the river and then up a bank which was so steep it was unlikely that any beast would come after it. I wondered whether this unusual behaviour was due to her having been as frightened by the buffalo as I had been.

By the beginning of August Elsa had become increasingly co-operative, but her son Jespah did not follow her example; every day he became more obstreperous. For instance, Elsa never interfered with our flock of goats, but Jespah now took much too much interest in them.

One evening when Nuru was herding them towards my truck, he made a beeline for them, rushed through the kitchen, passed within a few inches of the devout Ibrahim, who was kneeling on his mat absorbed in his evening prayers, dodged between the water containers and round the open fire and arrived at the truck just as the goats were about to enter it.

There was no doubt as to his intentions, so I ran and grabbed a stick, and holding it in front of him shouted, 'No, no,' in my most commanding voice.

Jespah looked puzzled, sniffed the stick and began spanking it playfully, which gave Nuru time to lift the goats into the truck. Then Jespah walked back with me to Elsa who had been watching the game. Often she helped me to control him, either by adding a cuffing to my 'noes' or by placing herself between the two of us. But I wondered how long it would be before, even with her support, my commands and my sticks failed to have any effect. Jespah was so full of life and curiosity and fun; he was a grand little wild lion, and a very fast-growing one too, and it was high time that we left him and his brother and sister to live a natural life. While I was thinking this, he was chasing after the other cubs, and in doing so tipped the water bowl over Elsa giving her a drenching. He got a clout for his pains and then she squashed him under her heavy, dripping body. It was a funny sight and we laughed but this was tactless and offended Elsa, who, after giving us a disapproving look, walked off followed by her two well-behaved cubs. Later she jumped on the roof of my Landrover and I went to make friends again and apologize.

The moon was full and in the sky the stars sparkled brilliantly, and Elsa, her great eyes nearly black owing to her widely dilated pupils, looked down at me with a serious expression as though saying: 'You spoilt my lesson.' For a long time I remained with her, stroking her soft silky head.

Suddenly we heard the whinnying and grunts of two love-making rhinos coming from the salt lick. Elsa glanced alertly towards the cubs, but when she saw that they were entirely absorbed in their meal, she decided to pay no attention to the love-sick pair and presently we heard them crossing the river.

George had now joined me and brought the anti-poaching team with him. This consists of a sergeant, a lorry driver and Game Scouts, all Africans. The team is sent wherever its services are most needed and therefore operates all over the North Frontier District. The first thing George wanted them to do was to find some man belonging to the tribe on the far side of the river who would be willing to supply information about the poachers and any other illegal activities which might endanger the lives of wild animals.

A most effective bush telegraph operates throughout the

N.F.D., based on the services of 'informers' who so far from feeling ashamed of their profession have come to regard themselves almost as auxiliaries of the Game Department. Indeed, informing is an accepted practice and without informers it would be impossible to control poaching over such a vast area. The informer is well rewarded for accurate information, because he incurs great risks.

Yet in spite of the services of these informers, poaching is difficult to put down, firstly because the culprits far from being deterred by the prison sentence rather welcome being provided with food, clothing and shelter in return for work which makes a break in the monotony of tribal life; secondly, because being recognized as a poacher is regarded as a proof of an enterprising and plucky nature.

Now the anti-poaching team was established there we had every intention of leaving Elsa and her family to look after themselves. Her wounds were more or less healed and we wanted the lions to lead a natural life. But when the Scouts returned we found that we had to change our plans. They brought in some prisoners and an informer told George that the poachers had determined to kill Elsa with poisoned arrows as soon as we left the camp. He also said that after burning the camp three of the culprits had climbed Elsa's big rock to hunt hyrax, but had given up when one of them got bitten by a snake.

When George interrogated the prisoners, he was warmly greeted by one who reminded him that it was fourteen years since he had first been sentenced by George for poaching. He added that he had had four sentences from him since then. This seemed to be the accepted basis of an old friendship.

We realized that as the drought increased, so would the poachers' activities, and however efficient the anti-poaching team might be, it would be impossible for them to prevent Elsa, if unfed by us, from hunting farther afield and risking an encounter with the tribesmen.

Obviously, if we stayed on, the cubs' education in wild life would be delayed and they would probably get spoilt, but it was better to face this than risk a tragedy.

One evening the tsetse flies were particularly active and Elsa and her two sons rolled on their backs inside my tent trying to squash their tormentors. In doing so they knocked down two camp-beds which were propped up against the wall. Elsa lay down on one of them and Jespah on the other, while Gopa had to be content with the groundsheet. The sight of two lions lolling in bed, while far from our ideal picture of Elsa's family returned to a wild life, was comic enough. Only Little Elsa stayed outside: she was as wild as ever and nothing would induce her to enter the tent, so she at least appeased my conscience.

Recently we had had a new visitor, a genet cat, who came every night to snatch a meal from the scraps which the lions had left. She was an attractive little beast, quite unafraid when George flashed a torch at her. Gradually she grew tamer and one night woke George up by upsetting some plates which clattered to the ground; he switched on his torch and saw her only a few feet away, unperturbed. She continued to eat the remains of his supper, bits of cheese and roast guinea fowl. As the days passed she got bolder and bolder, though she took care never to appear until the lions had finished their supper.

One afternoon when we were on the river-bank with Elsa and her cubs, I had a good chance of examining her wounds, and I found that although I had given her plenty of sulphanilamide they had not yet healed. I took the opportunity also to examine her teeth and saw that two of her canines were broken.

The hook-worm infection she had suffered from as a cub had left a groove round the edge of her teeth and the breaks had occurred along these indentations. These broken teeth would, I thought, hamper her when hunting even though her claws were her main weapons.

When it got dark we went back to the tents; all that evening Elsa was alert and restless and eventually she and the cubs disappeared into the bush.

About midnight I was woken up by the roaring of several lions. This was followed by the frightening noise of a fight and after a pause, another fight and later a third. Finally, I heard the whimpering of a lion who had obviously got hurt in the battle and I could only hope it was not Elsa. Next there was the sound of an animal crossing the river and then all was quiet.

At dawn, we got up and went out to track the spoors left by our quarrelsome visitors. We recognized those of the fierce lioness and her mate. Evidently Elsa had challenged them when they neared the camp. For six hours we followed her pug marks which led across the river to the Border Rocks; they joined up with those of the cubs.

All day we searched fruitlessly and at sunset fired a shot. After some time we heard Elsa calling from very far away, and eventually she appeared, followed by Jespah.

She was limping badly; but seemed to wish to get to us as fast as she could hobble, though she stopped once or twice and looked back, to see whether the other two cubs were coming. Both she and Jespah when they joined us showed how pleased they were by rubbing themselves against our legs. I then saw that Elsa had a deep gash in one of her front paws, which was bleeding and obviously causing her a lot of pain. The only way of helping her was to get her home and dress the wound.

The camp was far off, it was getting dark and judging by the many buffalo and rhino spoors we had seen, it was essential not to get benighted. Everything indicated that we should hurry, but in spite of George's impatient shouts urging us to make haste we had often to stop and wait for the little ones whose pace was rather slow. Jespah acted like a sheep dog running between George and the rear-guard trying to keep us all together.

For once, the tsetse flies were a help. Elsa was covered with them and so kept up with me in the hope that I would brush them off her back. Jespah, too, was attacked by them and, for the first time, pushed his silky body against my legs asking me to deliver him too from this plague. It was all against my principles to touch him, but it was difficult to resist brushing off the flies.

Elsa often stopped to spray her jets against a bush. Was she in love again?

We were all completely exhausted when we got back. Elsa refused to eat, but sat on the Landrover watching the cubs tearing at the meat, at intervals looking with great concentration into the darkness. It was barely nine when she left the camp with her family and about midnight we heard a lion calling from the Big Rock.

During the next days she came into camp every afternoon and I dressed her wounds.

When she was better, she and the cubs came along the river with us on a 'croc' hunt. Then we had another example of the way in which she could apparently order the cubs to stay put and be implicitly obeyed.

She scented a buck and stalked it unsuccessfully; meanwhile the cubs remained as still as though they had been frozen to the ground and there was never a question of their interfering with her hunt, though later they were lively enough splashing in the water and climbing trees. This they achieved by hooking their claws into the bark and pulling themselves up; sometimes they got as high as ten feet above the ground.

Another of Elsa's instinctive reactions showed up on this occasion. The cubs were playing within one hundred yards of a crocodile who lived in a deep pool but she plainly regarded this 'croc' as harmless. Perhaps she knew that he was replete, for she was quite unconcerned by his proximity though, as a rule, the slightest ripple on the water would cause her to become suspicious. We had always observed that she differentiated between harmless games, such as a tug-of-war between George and Jespah over a carcase and one that might become dangerous, or frightening, as when George threw a stick into the river. Then she would immediately place herself between the cubs and the water – either to prevent them from jumping into it or if they were alarmed perhaps to reassure them that the thing they saw was only a piece of floating wood and not the snout of a crocodile.

On the 12th August, the night before I had to go to Nairobi, Elsa and the cubs left the camp early and soon after they had gone we heard a lion roaring from across the river. Next morning George tracked the pug marks of this lion and close to them were those of Elsa and the cubs; the tracks led towards the Cave Rock.

Near there he observed vultures circling and when he reached the spot saw the remains of a rhino which had been killed several days earlier by poisoned arrows. The lion had evidently had a good meal off the carcase.

I returned on the 18th August and while we were having a belated supper, we heard two lions roaring. From the noise we gathered that they were approaching the camp rapidly from upstream. Elsa rushed off in their direction leaving the cubs behind; she returned after about three-quarters of an hour, but by then the cubs had gone, so she began to look for them all round the camp and seemed very nervous.

Suddenly we were startled by the most deafening roar which seemed to come from just behind the kitchen and George, looking in that direction, saw the torchlight reflected in the shining eyes of a lion.

Standing close to our tent Elsa roared back defiantly until, luckily, the cubs arrived. She took them off at once and soon we heard them hurriedly crossing the river.

After this all was quiet and we went to bed. But about 1.30 a.m. George was woken by a noise near his tent and flashing his torch saw a strange lioness sitting some thirty yards away. She got up slowly and he put a shot over her to speed her on her way, but this had no effect except to start another lion roaring.

For half an hour roars, growls and grunts succeeded each other, then the lions moved on.

Next evening Elsa came in very late and settled near the tents while Jespah, who was in one of his energetic moods, amused himself upsetting everything within reach; the tables were swished clear of bottles, plates and cutlery, the rifles were pulled out of their stands and the haversacks full of ammunition carried away, and cardboard containers were first proudly paraded in front of the other cubs and then torn to shreds. In the morning we found the family still in camp, a most unusual occurrence. The boys kept well inside the kitchen fence waiting for them to go, then, as they showed no intention of leaving, George walked up to Elsa, whereupon she knocked him down. After this George released me from my thorn enclosure and I tried my luck. I approached Elsa, calling to her, but as she looked at me through half-closed eyes, I kept on my guard while she came slowly towards me, and I was justified, for when she was within ten yards of me she charged at full speed, knocked me down, sat on me and then proceeded to lick me.

She was extremely friendly, so this, it seemed, was no more than her idea of a morning game. But she knew quite well that the knocking down trick was not popular with us and this was the first time since the birth of the cubs that she had indulged in it.

Later she took the cubs to a place below the studio, and in the afternoon we joined them there. Jespah was very much interested in George's rifle and tried his best to snatch it away from him, but soon he realized that it was impossible to do this so long as its owner was on his guard; after this discovery it was amusing to see how he tried to distract George's attention by pretending to chase his brother and sister. When George's suspicions were allayed and he put the rifle down to pick up his camera, Jespah pounced on it and straddled it. A real tug-of-war followed, which Elsa watched attentively. Finally, she came to George's rescue by sitting on her son and thereby forcing him to release his hold on the gun. She continued to sit on the cub for such a long time that I got quite worried about him. When she finally released him, though he looked longingly at the rifle and crouched near it, he was very subdued and left it alone. Nevertheless, for a while Elsa remained suspicious of his good behaviour and at intervals placed herself between him and the gun.

Finally, she rolled on her back with her paws in the air and moaned softly. The cubs responded at once and began suckling. Elsa looked utterly happy, but I could not help wondering how the cubs avoided hurting her with their sharp teeth. It was a most idyllic scene and just at that moment a paradise flycatcher flew over us trailing its white tail feathers like a long train behind it. The cubs were eight months old that day and she had every reason to be proud of them.

When they dozed off, their round bellies filled to bursting point, Elsa got up, arched her back, gave a long yawn, came over to me, licked me, sat beside me and rested her paw on my shoulder for some time, then she put her head on my lap and went to sleep. While she and the two small lions slept, Little Elsa kept guard over the family and twice unsuccessfully stalked a water buck.

When we were in bed we heard sounds of crunching, which went on until morning; evidently the family were spending the night in camp, finishing up the carcase. During the following day, they stayed very close to the tents. That evening we heard the cubs' father calling and thought it was because he was nearby that Elsa had preferred not to go far afield. For three more days she never left us.

11
Cubs and Cameras

There was truly a 'Garden of Eden' atmosphere about life around the precincts of the camp, for the animals who shared this territory with us had got so used to our presence that they often came very close without showing alarm.

There was the bush buck ram who came every day while we were having our lunch for his drink in the river opposite the studio. He browsed not only the greenery off the bushes but also a quantity of dried leaves off the ground, and sometimes he spent as long as an hour within sight of us and remained unconcerned even when we talked or moved about.

Then there was the water buck family, consisting of two males, three does and three youngsters, who, when they were together, would allow us to come quite close, but were much shyer if we came upon one of them when it had got separated from the herd.

The baboons were, of course, our oldest friends. Indeed, we had lived side by side for so long that we no longer paid any attention to each other unless something unusual happened. At this season the drought was so great that they started digging up the juicy roots of the reeds which grew on the rocks in the river. One old male was the pioneer in this enterprise. He took possession of a boulder on which a lot of reeds grew, he pulled up these and then began to dig energetically for their roots, often kneeling down to bite them off. Sometimes he had to knead the hard soil between his hands until it was loose enough for the roots to come out. Then he carefully peeled off their outer skins and stuffed them into his mouth, till his figure resembled a little barrel, and so he got his name. He was so accustomed to seeing us that when I filmed or sketched him from a distance of about twenty yards he scarcely bothered to glance at me. On this particular day, another male wished to join in the digging, and though much bigger than Barrel he was plainly afraid of him and waited for him to have his fill before intruding.

Then he advanced very cautiously, keeping a wary eye to see whether Barrel showed any sign of displeasure. When he thought all was safe, he jumped from the rock to the boulder and began digging opposite Barrel, who soon went to have a look at his work and decided to take over his pitch, whereupon the intruder meekly withdrew to Barrel's old site. Later, a still larger male came to investigate but got such a bad reception that he ran away screaming loudly.

Barrel was certainly a despot, he allowed no female to approach the larder, and so, like well-trained Victorian wives, they kept in the background, sitting on the bank in groups of five or six, suckling their babies, scratching each other's fur and seeking what nourishment they could get from the scanty grasses.

I spent three days sketching the males; they were easy to identify, Barrel by his personality, the first intruder by a deep scar on his nose and a third by a kink in his tail. A truce seemed to have been established between them, but the terms of the agreement evidently gave Barrel the right to take over any excavation if he chose to do so. When they had scratched their little island to rock level they abandoned it and went down river to some rocks close to a sandbank. Nearby lived a crocodile, which I knew well and had often unsuccessfully tried to shoot. Now I saw him stretched out to his full length of about eight feet only a short distance from the baboons.

I got my rifle and stalked him, but just as I came within range the baboons gave the alarm and when next day the same thing happened, I began to wonder whether they were acting as look-outs.

Many birds are used by animals as sentinels, and the giraffe often acts as a watch tower for zebra and antelope, but I was surprised that baboons, whose young are so easy a prey to crocodile, should help a 'croc'. Of course, fish provide a more easily obtainable meal and so perhaps these baboons knew that the 'croc' was replete.

A crocodile has as good a right to his life as any other beast, but he endangers the existence of all creatures in his neighbourhood and, in this particular case, my sympathy was on the side of the fish, because we had lately become friends.

It happened in this way. One morning when I was sitting in the studio I ate a banana and threw the skin into the pool below me. Immediately there was a commotion, and the silvery bodies of many fish twisted and leapt in their efforts to snatch it from each other, until one, with a swift jerk, grabbed it and dived under a rock with its prize.

I knew very little about fish and was surprised that a banana skin should prove such an attraction, but since, when I threw another into the pool, a fierce fight followed,

I realized that they were much valued. After this I tried the fish out with all sorts of food, except meat which we couldn't spare; bread, banana, paw-paw and mango skin were their favourites in that order. After a few days whenever the fish saw us on the river-bank they congregrated in large numbers and if I held the titbits under water they would take them out of my hand. They were such jolly fish that I was very unhappy when our supply of bread and fruit ran out.

We couldn't afford to feed them on meat because Elsa and the cubs needed such vast meals and these were often shared by uninvited predators of many different species. By day, when the meat was hung up in the shade, there was a mongoose which crept on to it from an overhanging branch, and the monitor who jumped up to it from below; and by night there were a couple of jackals and a genet cat and a civet cat, and hyenas.

Birds were as much a part of camp life as beasts. A pair of hammerkop storks whom we had known for many months, were frequent visitors to the studio. They lived close by and daily we used to see their top-heavy heads bobbing into the muddy pools left during the drought by the diminished river. Now they had suddenly taken to coming within a few yards of us. We thought this was because the buffalo, in spite of our encounter, was still about the place, and little puddles had arisen in the depressions made by his hoofs, in these the hammerkops evidently found tasty food.

They were not the only birds which visited us; there was also a splendid pair of Hadada ibis whose long wailing call had become a motif in our life. A less tame visitor was a large picturesque Goliath heron who frequented the rapids.

I never tired of watching all these creatures and every day brought its surprises.

Even now, as I am typing these words a troop of some fifty baboons are pacing along the bank opposite me. In the middle of them are three bush buck, a ram, a doe and their fawn. They seem to have joined the troop for safety and are not in the least concerned when a baboon brushes past them.

No scene could be more peaceful or further removed from the generally accepted picture of baboons tearing small animals to pieces. I thought that, if it were not threatened by the poachers, wild life here would be ideal, for even the fierce lioness is much less of a danger to Elsa than these men. In any case, she is a natural part of bush life; so are feuds between lions.

It was encouraging to know that Elsa now went out to meet her enemy. We had first noticed this during the third week of August, the night when Elsa and the cubs were eating their supper in front of the tent. Suddenly she growled and went off, and only returned an hour later. During that night I heard two lions approaching camp, and soon afterwards a fearful quarrel broke out. Towards dawn I heard Elsa moving the cubs in the direction of the Big Rock. In the afternoon we met her in the bush on her way to camp, her head, especially near her wounded ear, covered with bleeding bites.

When she reached home I got out the remains of their last night's supper; there was not much left. Elsa wouldn't touch it but the cubs ate ravenously. When a new carcase was brought by the boys she, too, began to eat. I wondered why, if she was so hungry, she had refrained from touching the first course I had provided. Could it have been that she saw that there was not enough to go round and wanted the cubs to have a chance of filling their bellies before she took her share?

That evening Ibrahim arrived with a new lion-proof Landrover I had recently ordered. He also brought the mail, and I settled down to read an article about Elsa in the *Illustrated London News*. She was described as a world-famous animal. This was gratifying, but at the moment poor Elsa was tilting her head in great pain.

When she joined us in the studio next day she was still very distressed, not that this prevented her from disciplining Jespah with a series of well-aimed clouts when, intrigued by the clatter of my typewriter, he teased me.

Poor Jespah, he still had a lot to learn, not about the wild life which is his, but about the strange world which is ours and which he showed so great a wish to investigate. One night, for instance, I heard him apparently very 'busy' in George's tent. How 'busy' I only discovered next morning when I noticed that my field-glasses were missing. Eventually, I found bits of their leather case in the bush below the tent. They bore the imprint of Jespah's milk teeth. Close by lay the glasses, and luckily, by some miracle, the lenses were intact. Yes, there was no doubt that Jespah could be a nuisance but he was irresistible and one couldn't be cross with him for long.

At eight months he had now lost his baby fluff but his coat was as soft as a rabbit's. He had begun to imitate his mother and to wish to be treated by us as she was. Sometimes he would come and lie under my hand, evidently expecting to be patted and, though it was against my principles, I occasionally did so. He often wanted to play with me, but though his intentions were entirely friendly I never felt sure that he might not bite or scratch me as he would his own family. He was not like Elsa who controlled her strength on such occasions, for he was much closer to a wild lion.

We were both very interested in observing the different relationships which Elsa's cubs were developing towards us. Jespah, prompted by an insatiable curiosity had overcome his earlier inhibitions, mixed with us and was most friendly, but allowed no familiarity.

Little Elsa was truly wild, snarled if we came close and then sneaked away. Though she was less boisterous than her brothers, she had a quiet and efficient way of getting what she wanted. Once I watched Jespah trying to drag a freshly killed goat into a bush. He pulled and tugged and somersaulted across it – but nothing would move the carcase. Then Gopa came to his aid and between the two they tried their best – but finally gave up exhausted and sat panting next to it. Now Little Elsa, who had watched their exertions, came along and pulling hard, straddled the heavy load into a safe place where she was joined at once by her panting brothers.

Gopa quite often made use of the tent when the tsetse were most active, and it was on these occasions that I noticed how jealous he was. For instance, if I sat near Elsa he would look long and scrutinizingly into my eyes with an expression of disapproval and made it extremely plain that she was his 'Mum' and that he would prefer me to leave her alone. One evening I was sitting at the entrance of the tent while he was in the annex at the far end and Elsa lay between us watching both of us. When Gopa started chewing at the tent-canvas, I said as firmly as I could, 'No, no'; to my surprise, he snarled at me, but stopped chewing. A little later he took up the canvas again and, though my No was answered with another snarl, he again stopped.

So far, all the cubs responded when we said No although we had never enforced our prohibitions with a stick or anything else which could frighten them.

After a peaceful day and night around the camp, Elsa and her cubs left early one morning and crossed the river, so I was surprised when shortly afterwards Makedde reported having found the spoor of a lioness which last night had come from upstream as far as the kitchen and returned the same way. Was this the fierce lioness? Though Elsa had shown no sign of alarm, she kept away for one and a half days and when she did return it was after dark. She kept the cubs hidden at some distance and dragged the meat away quickly, keeping out of view with the cubs all the time. Next morning all of them had crossed the river. A few nights later while the family were still in camp, we heard, towards dawn, two lions approaching from upstream. Elsa at once took her children away and I saw them in the dim light rushing towards the studio. Soon afterwards Elsa returned alone and trotted determinedly in the direction of the lions. Neither I nor the boys heard a sound, though we listened intently, until after half an hour Elsa came back and called her cubs. There was no reply and she rushed round desperately calling and calling. As soon as I had disentangled myself from my thorn enclosure, I joined her in her search, but she only snarled at me and, sniffing along the road, disappeared in the direction of the Big Rock. A little later we heard lots of 'whuffings' coming from this direction, but assuming that the two lions were close, we did not follow Elsa until the afternoon, when all was quiet. On the road we found not only Elsa's pug marks, but also those of another lioness, both leading to the rock.

Elsa did not come to camp that night, but two hours after George returned from Isiolo the next afternoon, Elsa arrived with her cubs, all fit but very nervous. She inspected the bush round the camp several times and left long before daylight.

By the beginning of September the drought was severe. Thanks to patrolling by the anti-poaching team few wild animals had been killed, but the team could not remain indefinitely in the area, since their services were urgently required in other parts of the country. When they left George would only have his small staff to depend on and there was no hope of the rains beginning till the end of October.

It was welcome news when we heard that Sir Julian Huxley was soon coming on a mission sponsored by UNESCO to investigate the problem of the conservation of wild life in East Africa. When he wrote asking us if we could show him parts of the Northern Frontier Province we were very pleased as this would give us the opportunity of acquainting him with the local problems, and the lack of means for dealing with them.

We believed that Sir Julian's visit would be a great encouragement to all those interested in the preservation of wild life. We also knew that he wished to see Elsa. We limited her visitors to those who had good and sufficient reasons for seeing her and, as Sir Julian clearly had those, we were glad that he should spare time to do so, but we were concerned to ensure that he ran no risk. We solved the problem by agreeing that none of the party, which included Lady Huxley, Major Grimwood the Chief Game Warden, and their pilot, should leave their car if Elsa did appear during the short time that they were to be in her area.

Between 7th and 9th September we showed Sir Julian something of the North Frontier District, and late one afternoon we arrived in Elsa's domain.

We fired the usual signal shots and twenty minutes later were delighted to hear the barking of baboons which usually heralded the arrival of Elsa and the cubs. In her enthusiastic welcome she nearly knocked me down and then hopped on to the top of the Landrover. Meanwhile, the cubs were busy dragging the carcase we had provided for them into a 'safe' place. For half an hour we watched them and then left. Elsa had a very puzzled expression when she saw the cars going off after such a short visit.

Next morning our guests flew back to Nairobi and we returned to camp. Three hours after our arrival Elsa and her cubs reappeared, all very heavy and lazy after having gorged on the meat we had left the day before. That night we heard two lions calling from far away, and at once Elsa and the cubs crossed the river; from the far side she kept up a conversation with the lions which went on long into the night. The next day while I was having tea in the studio, Elsa appeared, very wet and unaccompanied by the cubs, but later they came and we spent a very happy evening around the camp. Jespah behaved as though he owned the place and even lay down on George's camp-bed, which was vacant as he had gone back to Isiolo.

During the night I heard Elsa calling to the cubs and

walking in circles round the camp and the salt lick. I felt rather worried at hearing no cub noises and was still more worried when on the following afternoon she arrived alone from across the river. However, my fears were unnecessary for later they appeared and the next morning came in early asking for breakfast. After two hours of gorging they left and in the afternoon the Toto and I followed them and found the family resting on the Whuffing Rock. Elsa soon spotted us, came down and rubbed herself against me. Jespah came only a short way and then sat half-way down the rock watching us.

After we had returned to camp George arrived bringing a lorry as well as his car, and, attracted by the noise of the engines, Elsa and the cubs soon turned up. George told me that next morning David Attenborough and Jeff Mulligan were arriving from London and that we were to collect them at the nearest airstrip. For some time we had been corresponding with David Attenborough about making a film of Elsa and her cubs for the B.B.C.

We had had previous suggestions for filming her but these we had refused fearing that the arrival of a large film unit might upset her. The coming of only two people was much less worrying, but even they would need constant protection. We hope to provide for their safety at night by making one sleep in my lion-proof Landrover which was driven into a large thorn enclosure; our other guest's sleeping quarters were to be a tent rigged up on a lorry which also stood in the enclosure. Another tent would serve as dressing-room, bathroom, laboratory and equipment store.

Soon after we had gone to bed we heard a lion roaring upstream and observed that Elsa at once left the camp.

Next morning, the 13th September, George called me early to his tent and there I saw Elsa, in a terrible state, her head, chest, shoulders and paws covered with deep bleeding gashes. She appeared to be very weak and when I knelt beside her to examine her wounds, she only looked at me. We were very much surprised for we had not heard any growls during the night and were quite unaware that a fight had taken place. When I began to try to dress her wounds Elsa struggled to her feet and slowly dragged herself towards the river, obviously in great pain. I went at once to mix some M. and B. tablets with her food hoping to counter the risk of sepsis in this way, since any external treatment was obviously going to hurt and irritate her. When all was ready I spent twenty minutes looking for her but could find no trace of her. Then I had to start off to meet our guests, leaving George to search for the missing cubs. It was the worst moment to have visitors – let alone film producers, and I feared that they might have no chance of doing any work. I greeted them with this depressing news and soon realized that we had been more than lucky in finding two such animal lovers as David and Jeff.

We arrived in camp at lunch-time and found George who had just returned from a fruitless search for the cubs. While our guests settled in I went to look for Elsa and found her under a thick bush near the studio. She was breathing very fast and lay quite still as I swished the flies off her wounds. I went back to camp to get water and to mix the M. and B. tablets with her meat. When David saw my preparations he offered to help, and walked with me to the studio carrying the basin of water. I made him put it down a short distance from Elsa and then I took over.

Poor Elsa, I had never before seen her in so much pain. She made no effort to raise her head and it was only when I lifted it that she began to drink; then she lapped for a long time. After that she ate the meat but made it very plain that she did not want company, so we left her.

Since there was nothing more we could do for Elsa, George and I set out to look for the cubs on the other side of the river. We walked shouting all the names by which we address Elsa and also calling Jespah. Finally, behind a bush, we caught sight of one cub, but as we approached it bolted. In order not to frighten it further we decided to go home and hope that the cubs would make their own way back to their mother. Jespah was the first to do so; about six in the evening he crossed the river and rushed up to Elsa, then we heard another cub miaowing from the far bank. Elsa heard it too, and dragged herself to the river-bank and began calling to it. It was Gopa and when he saw his mother he swam across. I provided some meat which the little lions devoured, but Elsa would not touch it. While Jespah and Gopa were eating we took our guests for a stroll along the river and were much surprised on our return to find Elsa on the roof of the Landrover which was parked in front of our tents. We had our drinks and our supper within a few yards of her, but she took no notice of us. We remained anxious about Little Elsa until some time after we had gone to bed George spotted her coming into the camp.

Soon after midnight the family moved off and a little later we heard the roars of the fierce lioness. During the following day Elsa kept away, and we knew why, for George saw the fierce lioness on the Big Rock. That night we again heard her roaring. We were very worried about

Elsa, so, as soon as it was light, George went upriver to try to find her, while I went in the opposite direction accompanied by Makedde, Nuru and a Game Scout; we carried water with us in case we found her. We picked up Elsa's spoor half a mile beyond the Border Rock, which was farther than we had ever known her to go. I began calling and presently she came out from behind some rocks. She reconnoitred the neighbourhood to see whether all was safe and then the cubs appeared. They were terribly thirsty. I could not pour the water out quickly enough and I had some difficulty in avoiding getting scratched and in preventing the plastic water bowl from being torn out of my hands.

When we started for home and rejoined the boys who had stayed behind, both Elsa and Jespah sniffed very suspiciously at the Game Scout. He followed my advice and stood absolutely rigid but his face betrayed less ease than his action suggested. As soon as it was possible I sent him ahead back to camp with Makedde.

Elsa's wounds had improved but still needed dressing. It took a lot of coaxing to get the family to follow us and we made our way slowly back to camp. Nuru stayed with me as gun-bearer, but when I thought we were nearly home I told him to go on and warn David of our coming, so that he would be able to film the lions crossing the river. After he had gone I felt a little uneasy and then became really worried for I found that I had miscalculated the distance and had lost myself in the bush. By then it was midday and very hot and the lions stopped under every bush to pant in the shade. I knew that the best thing to do was to find the nearest lugga and follow it, for it must lead to the river from which I would be able to get my bearings. Fairly soon I came upon a narrow lugga and walked along between its steep banks. Elsa followed me and the cubs scampered along some way behind her. I had turned a bend when I suddenly found myself standing face to face with a rhino. There was no question of 'jumping nimbly aside and allowing the charging beast to pass' as one is supposed to do in such encounters, so I turned and ran back along my tracks just as fast as I could with the snorting creature puffing behind me. At last I saw a little gap in the bank and before I knew I had done it, I was up it and running into the bush. At this moment the rhino must have seen Elsa for it swerved abruptly, turned round and crashed up the opposite side. Elsa stood very still watching the pair of us. This was very lucky for me and I was extremely glad that she had not followed her usual habit of chasing any rhino she saw.

A few moments later I was greatly relieved to see Nuru coming towards me. I was going to thank him for running to my rescue, but before I had time to speak he told me that he, too, had met a rhino and been chased by it and that this was what had brought him to where I was. We had a good laugh over our frights and then, keeping close together, we went back to the camp.

We found it deserted, for when Makedde had come in and told them that I had found Elsa, George, David and Jeff had set out to help me. I sent a scout after them to tell them that we had all reached home safely. Meanwhile, Elsa and the cubs had a game in the river and got nice and cool after their long hot walk. Then they retired with a carcase into the bush and remained there till about midnight when they crossed over to the other side of the river.

Assuming that there would be no opportunity of filming

the lions till late the next day we spent the morning photographing hyrax on the rocks. We returned hot and exhausted to a belated lunch and then went down to the studio, where camp beds had been put out for us so that we could enjoy a siesta. The beds were set out in a row, mine was on the outside, David's in the middle and George's beyond his. Jeff was some way off loading the cameras. Soon I fell asleep but woke up very suddenly to find a wet Elsa sitting on top of me, licking me affectionately and keeping me a prisoner under her immense weight; simultaneously David took a leap over George and went to join Jeff. Between them they quickly got the cameras working. Elsa made a bound on to George, greeted him affectionately and then walked in a more dignified manner up to the tents and settled herself inside one of them. She completely ignored the presence of our guests and behaved in the same way later in the evening when we were having our drinks. She had been inside a tent with Jespah and coming out passed within six inches of Jeff's feet, but did not take the slightest notice of him; so far as she was concerned he might not have been there.

Next morning we followed her spoor and found her half-way up the Whuffing Rock sleeping. As we did not wish to disturb her, we went home and only came back after tea. This time we took with us a sufficient number of cameras to take films from every angle.

We were very lucky for she and the cubs could not have been more obliging and posed beautifully on the saddle of the rock. Finally, Elsa came down and this time she greeted all of us, including David and Jeff, by rubbing her head gently against our knees. She stayed with us until it got dark and we went back to camp, but the cubs, possibly made nervous by the presence of strangers, stayed on the rock.

Although Elsa had not seemed upset by being filmed I wondered whether she would come for her evening meal. Lately if even one of her favourite boys was visible she had kept away from the camp. I need not have worried; just as I was going to explain to our guests that she might very well not turn up I was nearly knocked over by her stormy greeting. The fact that she appeared confirmed my impression that while she has become much more nervous of Africans she does not seem in the least suspicious of Europeans.

I mixed a dish of her favourite meat with some cod liver oil and was taking it to her when Jespah ambushed me and licked the dish.

While this was happening Jeff was testing the sound recorder and happened to run through some recordings of the fierce lioness roaring. Jespah cocked his ears and tilted his head sideways as he listened attentively to the hated voice. Then he left his titbits and rushed to warn his mother of the danger.

On the following afternoon we again filmed Elsa on the rock and had further proof of her friendliness towards David and Jeff: this time she brought the cubs to play with us. I was most interested to observe that Jespah reacted just as Elsa used to when she was a cub; he knew at once whether someone liked him, felt a bit nervous of him or was really frightened, and treated him accordingly. David, I am sorry to say, he singled out for stalking and ambushing, and most of his time was spent trying to dodge Jespah. It was a great pity it was too dark to get a film of this game.

On their last evening our guests said good-bye to Elsa while she was sitting on the Landrover; they shook her paw and I felt that she had become more to them than a mere film attraction. I was most grateful to both David and Jeff for all the tact and kindness they had shown while making their film.

12

Skirmish with Poachers

On the afternoon of 21st September, George and I and the Toto met the family in the bush. Elsa greeted us as usual and Jespah licked both of us, but when he went on to lick the Toto Elsa stepped disapprovingly between the two of them. This confirmed that her attitude had altered for she is as fond of the Toto by now as she is of Nuru and Makedde; since the birth of the cubs she has objected when they approached Africans or vice versa. We were surprised, however, that the ban extended to the Toto.

During the night we heard a lion roaring and next day George found the spoor of a male lion near the camp. Later the Toto and I went in search of Elsa and found her on the Whuffing Rock but, though I called to her and, leaving the Toto at the base of the rock, climbed up to her, she paid no attention and did not even look at me, nor did the cubs. I stayed for about twenty minutes and then went home wondering whether Elsa's mate might be nearby and if this was the explanation of her behaviour. She did not come into camp that night but next afternoon we saw the family playing in the river. While the cubs splashed about and fought over floating sticks Elsa placed herself near to the Toto in a position from which she could keep an eye on all of us.

As we walked home Jespah became very much interested

in the Toto's rifle and persistently stalked and ambushed him. Elsa came to the rescue several times and sat on her son long enough to allow the Toto to get well ahead unmolested.

That evening the tsetse flies were particularly annoying and Elsa flung herself on the ground inside my tent miaowing for help in getting rid of them. I came in to perform my task but Jespah and Gopa had already rushed up to their mother and were rolling round squashing the flies. When I approached Elsa they snarled at me and when I began to deal with the tsetse she began licking the cubs no doubt to quiet their jealousy. Usually, I was allowed to do this for her, and any help was gratefully accepted. I was therefore surprised when next morning while I was watching the cubs enjoying a game with their mother Elsa spanked me twice and even jumped at me.

That night she only came into camp for a quick visit after we had gone to bed, and did not appear again till the following evening when she arrived with the cubs, but was very aloof, collected her meat, dragged it out of my sight and left soon afterwards.

When I returned the next evening from a walk on which I had found many fresh elephant spoors I saw Jespah busily reducing my only topee to pulp. This was a bore as I needed it when I went out in the hot sun. Elsa, perhaps to make up for her son's naughtiness, was particularly affectionate. We sat for a long time together near the river watching a kingfisher. It seemed to have no fear of either of us and came very close.

It was about this time that I began to notice how very jealous Gopa was growing, not only of me but also of his brother. When Jespah played with their mother he would push his way between the two of them and when Elsa came close to me he crouched and snarled until she went over to him.

After George went away I slept in the Landrover close to which the carcase was chained at night; by doing so I hoped to preserve it from chance predators. This arrangement did not make for undisturbed sleep, but it gave me a wonderful opportunity of seeing the nocturnal creatures of the bush.

I particularly liked a civet cat, an exceptionally dark-coloured and fierce one which used to take possession of the 'kill' to the exasperation of the circling jackals, but had only to raise its head for them to bolt away as fast as they could. If this was the civet cat which Jespah had once chased away I was impressed by his courage. So as to be able to watch this interesting animal, I used to allow it to have a good meal, but I was much less generous to the hyenas and the jackals.

One night I was woken up by the sound of breaking trees and the trumpeting of elephant. They were down by the river, between the studio and the tents, but gradually moved nearer which worried me, for I could not think what I should do if they came up to the tents. Elsa sat with her cubs by my 'sleeper' facing the noise and perhaps harbouring similar misgivings. We all listened intently. Suddenly I saw a huge shape moving along the top of the bank; it stopped and stood still for what seemed an endless time, then it vanished into the darkness. Elsa and the cubs kept as quiet as I did and remained in their 'guarding' position until the sound of crashing had ceased. Then I thought I saw her go off.

Soon afterwards my torchlight was reflected in a pair of

green eyes which gradually came closer. Assuming it was a prowling predator, I got out of the car intending to cover the carcase with thorns, but before I had dragged one big branch into position Elsa bounced on me. I climbed back into my bedroom, then when she and the cubs seemed to have finished their meal and gone away, I came out again for I was determined not to give a free meal to the jackals. Once more Elsa jumped on me and defended her 'kill'. We spent the rest of the night watching each other. She won the game but probably at the cost of eating a lot more than she wanted.

One morning in the last week of September a tribesman came to ask for help in chasing away two lions which had killed a water buck near his home. I sent out two Scouts with him. So far as they could discover from examining the spoor, the lions had spent one night at their kill and had afterwards gone ten miles away to a hill which was their usual lair.

I was rather pleased to know that the tribesmen should be aware that there were other lions in the neighbourhood as well as Elsa, so that they would not necessarily blame her if their livestock were attacked.

By October both Billy Collins and I felt that it would be useful to meet and discuss plans for the sequel to *Born Free*.

I went to Nairobi to meet him and on our drive back was happy to find that he did not seem to have developed any resentment against Elsa or fear of her in spite of her peculiar behaviour towards him during his last visit. I had hoped to arrive in camp before she turned up, but in fact we did not get in till supper-time and found the family in front of the tents eating. I was a little apprehensive, but

Elsa welcomed us both in the most friendly fashion and then returned to her dinner. We spent the rest of the evening within a few yards of her but she paid no attention to us.

George told us that during the nights of 7th and 8th October a lion had roared close to the camp; that night as soon as we had gone to bed Elsa left the camp, perhaps to join him.

It was very hot indeed and the bush was depressingly dry, so that even the studio, which is usually cool, was oppressive when we went there next morning and started our work. Although we were much distracted by baboons, antelopes and various birds, we achieved a lot and it was not till after tea that we went to look for Elsa. We did not find her on our way out, but as we were returning to camp along a little game path I suddenly felt her and Jespah rubbing themselves against my legs.

Elsa treated Billy just as she did us, but Jespah was greatly intrigued by his white socks and tennis shoes. Crouching low and hiding behind every available tuft he prepared to ambush him, but we intervened, so eventually he became disgusted at being thwarted and went off and joined the cubs. Elsa spent the evening on the roof of the Landrover.

Next morning she woke me up by licking me through my torn mosquito net. How had she got into my tent? I was worried in case she might also have tried to visit Billy, and shouted to him. He replied that Elsa had only just left him. At this moment the Toto arrived with my morning tea. Seeing him, Elsa stepped slowly off my bed and moved to the wicker gate of the thorn enclosure. There she waited until the Toto pushed it aside for her, then she walked out sedately, collected the cubs and they trotted off towards the big rocks.

I dressed quickly and went with some apprehension to find out how Billy had fared. When I saw him grinning at me from inside the wired sleeper I felt better. He told me that Elsa had squeezed her way through the wicker gate of his enclosure which we had barricaded with thorn and had then jumped on to the Landrover. Only when she realized that she couldn't get at him, had she gone off to visit me.

She had never paid the slightest attention to David Attenborough or to Jeff who had slept in the same position. The only people whose beds she insisted on sharing were George's and mine, so I interpreted her behaviour towards Billy as a great compliment. I don't know whether he felt the same about it.

In the afternoon we visited the family which we found on the Whuffing Rock. As soon as Elsa and Jespah spotted us they came down and gave us a great welcome. Makedde was with us and Elsa greeted him too, but by stepping briskly between them she pointedly prevented Jespah from rubbing his head against Makedde's legs. Gopa and Little Elsa stayed on the rock, but after we had walked some hundred yards into the bush Elsa called them and they came down, but kept out of our sight. Only when we reached the river did they appear, and then they behaved very quietly, sitting in the water to cool themselves while watching us attentively. Jespah later joined Elsa and was very affectionate but on our way home his antics delayed us till we became benighted. Although Billy had discarded his white socks he still fascinated Jespah who sat himself squarely in front of his feet, looking up at him with the

most cheeky expression and making all progress impossible. Billy tried to make a series of detours to avoid him
but in vain for the next moment Jespah was always at his
feet. Elsa intervened once or twice and rolled her son over,
but this only encouraged him to be more mischievous.
George had gone ahead but suddenly felt himself clasped
from behind by two paws and nearly tumbled over. Jespah
had a good evening's fun! It was only when we reached
camp and he settled to his dinner that we were left in
peace.

The 12th October was the last day which Billy was to
spend in camp, so we made a determined effort to find the
family but failed and had to make do with watching water
buck and bush buck drinking in the river while the hyrax
sunned themselves on the rocks. Later when the light
began to fade we were able to observe the tiny Bush-babies
leaving their thorny homes and beginning their nocturnal
life.

On our return we found Elsa and Jespah in camp. Billy
patted Elsa as she lay on the Landrover and stroked her
head, something which as a rule she only allows me to do.

Before he left Kenya we decided to show him the Tana,
which meant a detour of fifty miles, but we thought he
would enjoy it. On our way we passed a baobab tree
which is the largest in the area and according to its size
may be as much as eight hundred years old. It has two big
openings high enough above the ground to provide a safe
refuge; they lead into a 'cave' which could hold eight or
ten people. When George first saw this tree it was being
used as a poachers' hideout. He put a stop to that, but
there are still a lot of wooden pegs in its walls, which must
have served either as ladders or to hang things on. The

openings have at some time been artificially enlarged and the texture of their surface suggests that this was done a long time ago, perhaps two or three hundred years. It is curious that most of the baobabs around these parts have such holes. Elsa found them intriguing and always made a point of investigating them.

Farther on our way we met Grevy zebra, giraffe, gerenuk and water buck. We did not expect to see very much game in the bush for at this time of the year the drought is so intense that most wild life is concentrated round the rivers. Eventually we came to the Tana, the largest river in Kenya. We reached it at the point at which one of the four streams which we have to cross on our road to Isiolo joins it. We could usually count on seeing hippo here and this time we were not disappointed. Eight were in the middle of the six-hundred-yard stretch of water; wallowing and yawning their massive glistening bodies looking enviably cool.

Suddenly we heard dogs barking, and I saw George pick up a rifle and rush off in the direction from which the noise came. He plunged through the in-flowing stream and disappeared into a doam palm thicket; simultaneously two water buck jumped into the river followed by yelping dogs who soon overtook the swimming antelope and fixed their fangs into the backs and throats of their victims. The smaller buck had three dogs hanging on to it and was struggling desperately. Next we heard a shot and a dog fell back into the river. At this moment an African surfaced. But when he caught sight of Billy and me standing on the bank he dived back into the water. I was very worried as the Tana is not only infested by crocodile but there are several dangerous whirlpools at this spot and, besides these, there were the eight hippo lying straight across the poacher's course. George, too, must have realized this, for he put several bullets on the far side of the swimmer to indicate to him that he had better turn back. But undeterred by bullets, hippos, 'crocs' and whirlpools the poacher swam on, determined to make his escape. The hippos had submerged and we expected a tragedy to happen every moment, but the man following the water buck and the dogs made his way to the far bank and landed unharmed. Having done so he at once disappeared into the bush. Now another poacher appeared and with him more dogs pursuing a small antelope who sought the safety of the river in vain, for one dog got hold of its muzzle and tried to suffocate it while another hung on to its back. George shot both these dogs and the antelope bravely swam on for a hundred yards, but then sank and drowned.

While this was going on I ran to George with more ammunition. He told me that he had nearly trodden on one of the poachers who, taking in the situation, had leapt into the river. George had then put some shots over him to induce him to surrender but he had swum under the water round the river bend; it was this man who had been horrified when he surfaced to find Billy and me standing on the bank.

By now the Tana flowed peacefully carrying with it the bodies of victims and aggressors both sacrificed to man's hunting instinct – or perhaps, his greed for meat or money. I was sad that this tragic episode should be Billy's last impression of Kenya. It was typical of many unrecorded incidents and an illustration of the need to put down poaching if wild life is to be preserved in East Africa.

It was encouraging to know that educated Africans were

beginning to appreciate the value of wild life as a national economic asset. It is only with the understanding and co-operation of the Africans themselves that the last remaining strongholds of the larger fauna can be saved from destruction from which there is no recovery.

13
Tam-tam

During the second week of October, George returned to camp and for several days life went on uneventfully until one night the fierce lioness and her mate announced their arrival by impressive roarings from the Big Rock. Elsa took the hint and at once moved her family across the river.

Early next morning George saw the fierce lioness standing on the Big Rock clearly outlined against the sky. She allowed him to come within four hundred yards of her and then made off.

Elsa came in for a quick meal that evening but did not reappear for forty-eight hours. During this time we changed guard. Worried by Elsa's absence, I went out to look for her but could find no pug marks. Next morning we found her spoor and those of the cubs all over the camp, and I thought it very strange that they had made no sound to indicate their presence. Following the pug marks we found them mixed up with the imprints of rhino and elephant.

That evening the family turned up, but Elsa was in a queer mood; she showed no interest in me or in Gopa or Little Elsa and was entirely absorbed in Jespah. I felt really sorry for Gopa who tried very hard to attract her attention, rolling invitingly on his back with outstretched paws whenever his mother passed close to him, with no result except that she stepped over him to join Jespah.

About 8.30 p.m. two lions started roaring; all the family listened intently, but only Elsa and Jespah trotted quickly towards the studio, Gopa and Little Elsa after going a short way with them came back to finish their meal. They went on gorging until there was a frightening roaring so close that they rushed at full speed after their mother who by now had crossed the river.

I brought the remains of their meal into safety, which was as well, for the lion duet went on all through the night. The following afternoon when the light was already fading Makedde and I saw a lioness climbing up the Big Rock and then sitting on top of it – undoubtedly this was the fierce lioness. I got out my field-glasses and had my first good look at her. She was much darker and heavier than Elsa and rather ugly. I observed that she was staring at us. Suddenly there was a scream close to us and the next moment the bush seemed to be alive with elephants. Makedde and I ran back to camp as fast as we could. All that evening the elephants trumpeted and rumbled as they went down to the river to drink. Besides this the lioness kept on roaring from the top of the rock. There was no question of sleep that night and Elsa naturally kept away.

In the morning we tracked the fierce lioness's pug marks and those of her mate; they had gone upstream back to the area in which we believed they usually lived. Elsa no doubt knew this for that night she brought the family into camp for their dinner. She now paid little attention to me until the cubs had settled down to their meal, then she was as affectionate as ever. This was plainly a new strategem she had devised so as not to arouse their jealousy.

The air was oppressive and lightning streaked the horizon at frequent intervals; soon after I had gone to bed a strong wind started blowing, the trees creaked and the canvas of the tent flapped; then the first drops of rain fell and it was not long before I seemed to be under a water spout. The downpour continued throughout the night. We had not expected this deluge and had not hammered our tent pegs in; as a result the poles collapsed and I spent my time trying to raise them sufficiently to keep some shelter over my head, while a river seemed to run round my feet.

When at last the freezing hours came to an end with daybreak, I looked forward to a cup of hot tea to warm me up but none appeared, for the firewood was too wet to kindle and besides, the boys had spent the night in the same conditions as myself.

When I emerged I saw that George's tent had also collapsed and from inside it I heard Elsa moaning in a low voice. Soon she appeared with Jespah and Gopa, rather bedraggled but dry. But even this downpour had not induced Little Elsa to seek shelter and when I caught sight of her outside the thorn fence I saw that she was drenched.

I began to sort out our soaked belongings and remove them to the cars to save them from the lions, and in this I was 'helped' by Jespah who had great fun defending each box I wanted to move. When I had finished my work Elsa, Jespah, Gopa and I crowded into my tent and Little Elsa consented to come inside the flaps but no farther; at least she had some protection there.

The rain continued for four days with only short respites in the late afternoons; visibility was reduced to a few yards. This was nothing unusual, for the rains vary a great deal in this part of Kenya. A hilly area may have a rainfall of a hundred inches in the year, while the surrounding plains record only fifteen inches.

Elsa's home though in semi-desert country benefits from a nearby mountain range from which several small streams run into the arid region. The one nearest to the camp now rose higher than I had ever seen it. A roaring, red torrent thundered over its banks and flooded the studio up to the level of the table, depositing a great deal of debris including a doam palm which had been uprooted. I was exceedingly glad that Elsa and the cubs were on our side of the river and that we had sufficient food for them.

Within three days the scorched parched surroundings of the camp had become green and the dry brittle bush had turned into luxurious vegetation. But it seemed as though it had exhausted its strength in putting out such a profusion of many-coloured flowers, for within three or four days the ground was carpeted with many-coloured petals.

The animals of the bush reacted instantly to the change from the barrenness of the drought to the rich abundance which succeeded it. On the day that the rains broke I watched the weaver birds return and set to work above our tents building nests and patching up the old ones. Through the drumming of the rain I could hear them twittering, and they appeared to be quite unperturbed by the downpour. They finished their nests in a matter of two or three days. A week later I found the first pale turquoise-coloured eggshells and for two or three weeks afterwards the ground was littered with them. These blue-green eggs contrasted strikingly with the multitude of scarlet insects which had suddenly emerged from holes in the wet sand. They always came out immediately after the first rains but vanished in a few days' time: now they were everywhere and looked like rolling velvet beans.

It was exactly one month after the rains had started and the yellow black-headed weaver colony had arrived above our tents, that I picked up a fledgling which had fallen from its nest. It was naked except for a few feathers which, as they were still encased in their sheaths all but a tiny bit of fluff at the end, looked more like quills. When I held it in my hand to keep it warm the fledgling looked pathetically defenceless. But however frail and helpless, it was possessed by a strong instinct for survival and never stopped crying for food. Although our staff spent most of their time catching grasshoppers, there were never enough to satisfy the hungry bird. I tried unsuccessfully to place it inside a nest which I hung near to those of other weavers in the hope that they might adopt the orphan. Every two hours I gave it the catch which usually amounted to about twenty grasshoppers' abdomens which I placed with forceps in the fledgling's throat. It thrived on this diet, and on the second day already welcomed my approach with loud chirpings and stretched its bald head as far as it could out of the entrance hole. I kept the nest in its natural position with the funnel-like entrance pointing to the ground so that the occupant was not only protected against rain but could also keep the nest, which was lined with soft guinea fowl feathers from our kitchen rubbish, clean of excrement. The little bird's instinct for cleanliness was remarkable and every time it had to empty its intestines, it climbed to the edge of the nest, let its tail overhang the entrance and carefully let its droppings fall to the ground below. Even when I held it in my hand, it always warned me if it were going to excrete by its restlessness until it found a finger on which to perch and from which it could relieve itself without soiling my hand.

George noticed that every night the weavers seemed to give vent to this urge in unison. He was often woken up by their sudden sleepy twitters which were succeeded by a sound like raindrops caused by their droppings falling on the tent canvas. This lasted for a few moments after which all was quiet again. As this happened two or three times during the night it seemed obvious that the weavers were gregarious even in this habit.

I called the little fledgling Tam-tam which in Swahili means candy or sweet. During the night it slept inside its nest which I placed on top of my mosquito net, where it was not only sheltered against the rain but where I could also scratch it from underneath if the little one cried. Even during the night Tam-tam kept her home clean and each morning I found the droppings neatly lined outside the nest. As soon as the first weaver awoke Tam-tam responded and from then on I had no peace. No scratching from below would make her forget her hunger. But at this early hour no insects could be found for the grass was still heavy with dew. One day I tried to help by giving Tam-tam the yolk of a hard-boiled egg which she ate greedily. An hour later to my horror I found the yolk bulging in a lump underneath the transparent skin above the neck. There was no question of a pouch. I tried to massage it away without success. To watch what happened, I kept Tam-tam in my hand for the next two hours by which time the yolk had dissolved. Apart from slight constipation which I lubricated with fat grasshoppers, Tam-tam was none the worse.

Just before dark she always seemed especially hungry,

but this also was a bad time for harvesting grasshoppers as the lions were in the camp. The feeding problem solved itself one late afternoon when Elsa flung herself down in the tent asking me to help her with the tsetse flies. I had Tam-tam in my hand and no chance to place her safely elsewhere. So keeping her hidden in my hand, I caught the tsetse off Elsa's back with the other; then it occurred to me that the tsetse might provide an ample food supply for Tam-tam's needs. She took them so greedily that I collected a good supply for next morning's breakfast.

Within the next three days Tam-tam developed feathers which showed she was a female. I watched her naked under-parts growing the softest fluff in one single day and the yellow tissue lining the beak being reduced to small spots at the corners.

On the fourth day after I had rescued her I hung the nest with her inside it on a branch of my thorn enclosure to let her enjoy the morning sun. As usual, she chirped as loudly as she could asking for food. This attracted other weavers and within a few minutes some twenty-five females and about five males surrounded the nest. Finally, a female went inside and remained there for about two minutes. As I did not know whether she was hostile or friendly I placed Tam-tam outside the nest so as to watch what went on. Immediately Tam-tam hopped from one branch to the next until she landed in the high grass. Here some female weavers guided her through this jungle to the river bush.

I did not think Tam-tam could fly well enough to be safe from kites of which there were several around the camp. I also felt worried about snakes as George had shot a cobra near the tent the day before – it was obviously looking for fallen fledglings – so I collected little Tam-tam and kept her well fed in her nest close to my table in the studio. She knew her name by now and whenever I called she appeared in the entrance hole chirping excitedly and doing an agitated shivering dance. I took her many times into my hand but she never ventured farther than on to the table or the typewriter.

Next day she was in the studio with us when suddenly she flew from her nest and disappeared into the surrounding bush. Although she replied to my calls and stayed close to where I was working, she kept out of reach. We hoped that hunger would soon bring her back, but although she repeated her dances and calls more frequently as the day advanced she was perhaps still too young to realize that she had to come to me for food, and expected me to bring it to her as her mother would have done. I became very anxious about Tam-tam's safety when in the late afternoon Elsa and her cubs appeared and made it difficult for us to catch her; and by the time we succeeded in coaxing the lion family away and settled them at their dinner near the tent, the light was fading rapidly.

By now Tam-tam had perched on the topmost branches of a bush surrounded by thick undergrowth far out of our reach. I was desperate for soon it would be dark and the fledgling might fall an easy prey to nocturnal enemies. We started chopping down the undergrowth so as to reach her. It was surprising that in spite of the noise of the wood cutting and the bending down of the branch on which she roosted she did not fly away, and waited until I could take her gently into my hand. When I finally settled with her in the tent, feeding her with the tsetse flies off Elsa's back, it was a strange sensation to feel this nearly weightless little bird quivering in my hand, her tiny heart beating under the softest fluff, while I sat close to Elsa stroking her with my other hand and feeling her affectionate response. I had become very attached to little Tam-tam, but how long would she consent to stay with me? Within a few yards was a colony of hundreds of busy, chattering, happy weavers; she belonged to them and only accident had put her into our care.

After giving her a generous breakfast of tsetse flies I again placed her inside her nest in the sun. She was immediately joined by two female birds who went in turns inside the nest. Soon afterwards Tam-tam emerged and flew in a long swoop towards the river bush, while both females kept close to her. For the next hour we watched these three flying from tree to tree, always staying within the colony and surrounded by other weavers. Sometimes one of the adult birds would go in search of food and return with an insect for Tam-tam and once we saw her being pecked by one of her protectors. We could easily recognize Tam-tam by her size and stumpy tail for she was the only youngster among the adult birds. Where, we wondered, were the other fledglings of the colony? Were they kept safely in the nests until they could fend for themselves? As the two female birds never left Tam-tam we could do nothing but leave her in their company. When we tried to find her towards dusk there was no sign of her, and we could only hope that she was safely tucked away inside a nest by her two foster-mothers and that they would take care of her.

After a week when the rains stopped I observed many baby animals: some small brightly coloured monitors were sunning themselves along the river but dived into the foaming waters when I approached them. Two tiny turtles, no larger than a shilling piece, were swimming near the studio. They were perfect miniature replicas of the adult turtles, about the size of a large soup plate, which I had often watched on the rocks opposite. But the queerest nursery of all I discovered one morning when I was walking down the river. Close to one of Elsa's favourite crossing places is a deep pool, where I observed what seemed to be gigantic tadpoles; they kept in a vertical position by paddling energetically. When I looked at them closely I saw that they were baby 'crocs,' though they must have measured no more than seven inches and could not have been more than two or three days old. They kept close to the steep bank and sometimes climbed up it. With their muddy colouring and large black blotches they were perfectly camouflaged. We counted nine of them swimming close together within about a square yard. One, however, seemed to act as sentry; he sometimes ventured on short excursions into the river, but always returned very quickly. Their heads were disproportionately large for their bodies and when in the water they supported them on floating reeds whenever they could and kept themselves afloat by energetic water treading. Their hind legs were broad and shapeless compared to those of an adult 'croc,' and their most striking feature was their eyes. These were the size of big peas and pale ochre in colour. They seemed blurred but the narrow vertical slit of the pupil was already clearly visible in some of them though not easy to recognize in others.

The enormous brows which protected their eyes and the very enlarged knob of their noses gave these 'crocklings' a grotesque appearance. There was no doubt that in spite of the blurring of their eyes their sight was very keen for, though we were six or seven yards from them, they dived at once if we made the slightest movement. On the other hand, when we talked or manipulated our cameras they paid no attention.

We went back to camp and collected some meat which we tied to a stick; this we placed invitingly close to the 'crocklings,' but it did not seem to interest them, nor did they pay any attention to some worms which we threw amongst them or to the water beetles, dragonflies and tadpoles which came into their orbit. However, when George started making his crocodile noise at them, '*imn, imn*,' they at once congregated together and turned their heads in the direction of the sound but did not approach and kept within the safety of the reeds. This at any rate proved that they were not deaf, so they must presumably have heard us speaking and clicking our cameras; but these sounds apparently meant nothing to them. We tried to find the broken eggshells from which they had hatched, but failed to do so; perhaps they were born on the other side of the river, which we could not cross on account of the flood. Two days later we returned to the place and saw only a couple of baby 'crocs' and on our next visit there was only one left.

George had reached camp as soon as the condition of the ground made it possible for him to travel, and had brought five Game Scouts with him. They were to provide a permanent patrol and put down poaching. It was necessary that they should live some distance away from Elsa and from our camp, and so George now began supervising the establishment of their post and cutting a motor track to it.

In two weeks' time we hoped that this work would be

well advanced, then we would start deserting Elsa for increasingly long periods so as to compel the cubs to go hunting with her and assume their true wild life. Our unexpectedly prolonged stay in the bush had caused them to get a little too used to camp life, and, though we had no control over them, Jespah was now on quite intimate terms with us; but apart from this their wild instincts were intact and certainly Gopa and Little Elsa only put up with us because they saw that their mother insisted that we were friends.

We wondered whether she communicated her wish that they should not hurt us, which they were now well equipped to do, or whether they simply followed her example. Jespah in particular, when he was playing with us or when he was jealous, could have done a lot of damage if he had not controlled himself, but he always did so and even when he was in a temper gave us good warning of the fact.

Gopa was less friendly but so long as we left him alone did nothing to provoke an incident.

Little Elsa remained shy, though she now seemed less nervous of us than she used to be. We were surprised that none of the cubs ever attempted to follow Elsa on to the roof of the Landrover, though they often gazed up at their mother with disappointed expressions, when she was resting on the canvas to escape their teasing. Judging by their ability to climb trees they could very easily have jumped on to the bonnet and then taken another leap on to the roof, and indeed Elsa had done this at a younger age, but for some reason they seemed to regard the Land-rover as out of bounds.

During George's absence Jespah and Gopa used his tent

as a sort of 'den.' As a result on his return he found it rather crowded at night. I was a little worried; George prefers to sleep on a low hounsfield bed and with Elsa, Jespah and Gopa around it I wondered whether one night there might not be trouble, but they behaved remarkably well. Whenever Jespah tried to play with his toes, George's authoritative 'no' made him stop at once.

The extent to which they felt at home was illustrated when one night Elsa rolled round and tipped over George's bed, throwing him on top of Jespah. No commotion followed and Gopa who was sleeping near George's head did not even move.

On another night when the family were sleeping in the tent a lion started calling from the far bank and Elsa at once took the family off. We wondered whether it might have been the fierce lioness, for next evening they dragged their dinner between the tent ropes and the outer fly, ate it and finally buried the stomach there, which was not very pleasant for George. Soon afterwards we heard roars and Elsa crossed the river with the cubs. The water was still very deep but next morning we found the explanation of their daring swim when we saw the pug marks of a single lioness close to the camp.

A day later when we were returning to camp we found the family except for Jespah gorging on a carcase. It was not long before we discovered the missing cub behind the tents enjoying a roast guinea fowl which he had stolen off the table, but he had such a mischievous expression that we could do nothing but laugh at the little rascal. We were surprised, however, that he preferred cooked meat to fresh; next day we had a further surprise when we came across the family in the bush and found the cubs being suckled.

They were now ten and a half months old, and I do not think that they could have got much milk as Elsa's teats seemed to be empty.

Although they were still being suckled we now noticed the first signs of adolescence in Jespah and Gopa; they had grown fine fluff round their faces and necks, and if they looked a bit unshaven their appearance was certainly very endearing. Elsa greeted us warmly and while she was doing so, Jespah pushed himself between us and demanded to be patted too. Elsa watched us and then licked her son approvingly.

We walked back to camp together. In front of it were the remains of last evening's meal, but Elsa refused even to sniff at it and demanded a new 'kill.' Later a leopard grunted from the other side of the river, and this caused her to rush off leaving the cubs – after about fifteen minutes they followed her. We were very glad to see that Elsa now took the initiative and was prepared to defend her territory.

That night a lion roared and when we later traced his pug marks they led to the Big Rock; evidently something had given the cubs a fright, for on the 24th November, when Elsa swam over, they refused to follow her and she had to go back twice to encourage them before they, too, swam across. Once landed they had a great game, Elsa rolling Jespah round and round like a bundle, which he loved, and poor Gopa jumping clumsily between them asking to be noticed; when I came close to photograph them Gopa growled at me, whereupon Jespah gave him such a clout that he looked quite stupefied by his punishment. It was all done in fun, but it showed up the different characters of the brothers. But as always when they settled down to their dinner all jealousy was forgotten.

George had shot a guinea fowl and I brought it out hidden behind my back because I wanted to give it to Little Elsa. I waited for a moment in which only she was looking up and then showed it to her. She took in the situation at once and while continuing to eat with her brothers watched me carefully as I walked a little distance away. I waited until Jespah and Gopa were concentrating upon the meat and when only Little Elsa saw what I was doing, dropped the bird behind a bush. Then, when she alone was watching me, I kept on pointing from her to the guinea fowl until suddenly she rushed like a streak of lightning, seized the bird and took it into a thicket where she could eat it unmolested by the others.

Next day we saw the family sitting on the rocky platform on the opposite side of the river to the studio, below which there is a deep pool which was at one time inhabited by a large crocodile. The cubs seemed nervous and only Elsa swam across. We had brought a carcase with us, she grabbed it, and crossed the river with it, but this time avoided the pool and swam higher upstream where the bank was much steeper but where we had never seen 'crocs.'

The family were not apparently hungry, for they did not eat but indulged in a game of tree climbing; the cubs balanced on the sloping branches which overhung the river and seemed intent on tripping one another up and throwing their adversary into the water. Finally, Elsa joined them; she seemed to us to be giving them a demonstration of how to turn on a branch and how to go from one branch to another.

When it grew dark the meat was still untouched and as we neither wished to lose it nor to provoke a fight between Elsa and some chance predator George determined to recover it.

The first thing was to get the family over to our side, otherwise they would object to the removal of their 'kill.' While George went up the river out of their sight and began to wade across, I swung a guinea fowl temptingly in the air. This did the trick and brought the lions over to join me. Unfortunately, when George reached the carcase Elsa observed this, swam hurriedly back and defended it. It took a lot of coaxing on his part to let her allow him to float the 'kill' over, and even then she swam beside him with a very suspicious expression on her face. While this was going on the cubs rushed up and down the bank, obviously most upset but making no attempt to join Elsa. I was surprised for usually they showed no fear of the river and by now it was quite fordable. However, later that day they redeemed their reputation: shortly after dark, when we heard sounds which indicated that a rhino was at the salt lick, Elsa dashed after it and the cubs with her and judging by the snortings that followed the rhino must have made a very rapid retreat.

Brave the cubs certainly were to tackle such a great and fierce beast.

Jespah in his playful moods liked acting the clown. One day when he was being especially lively, teasing everybody and asking for a game, I placed a round wooden tea tray in a branch that hangs over the river to see what he would do about it. He climbed up and tried to grip the inch-thick rim between his teeth, using one paw to steady it as it swayed. When he got a sufficiently good grip to carry it horizontally he came down very cautiously, pausing several times to make sure that we were watching him. Finally, he reached the ground and then paraded round with his trophy, until Little Elsa and Gopa chased him and put an end to his performance.

George's leave was coming to an end and this seemed to be the right time for us to leave the camp. Elsa had by now got the upper hand of the fierce lioness and was able to defend her territory; the poachers seemed to have left the district and we hoped that they would not return at least until the next drought, by which time the Game Scouts would be able to deal with them, as their post was nearly completed and their patrols were already in action along the river.

Besides, the cubs were now powerful young lions, and it was time that they should hunt with their mother and live their natural life; also as they were growing increasingly jealous we considered that it would be unfair to provoke them by our affection for their mother into doing something which might be harmful.

We decided to space our absences. On the first occasion we had intended to leave for only six days, but in fact, because of very heavy rains, it was nine before I could return. I came alone and greatly missed George's help when I found myself obliged to dig the lorry and the Landrover out of the bog, a task that occupied us for two days.

Elsa did not turn up in answer to the shots we fired nor were there any signs of spoor around the camp, but these might well have been washed away by the flooding river. After a while, I walked towards the Big Rock and came upon Elsa trotting along with the cubs; they were panting and had probably come a long way in answer to my signal. They were delighted to see me and Jespah struggled to get between Elsa and myself so as to receive his share of the welcome. Gopa and Little Elsa, however, kept their distance. All were in excellent condition and as fat as they had been when we left. Elsa had a few bites on her chin and neck but nothing serious. Gopa had grown a much longer and darker mane than Jespah, whose colouring was very light in comparison to his brother's. In a year's time, I thought, what a handsome pride they would make, with two slender graceful lionesses, accompanied by one blond and one dark lion.

I had brought a carcase, but though Elsa settled down to it the cubs were in no hurry to eat and played about for some time before joining her. When she had had her fill she came over to me and was very affectionate and as the cubs were too busy eating to notice this there were no demonstrations of jealousy, which seemed to be what their mother had intended.

How anxious Elsa was to prevent rows or ill-feeling was clearly shown next day. I had given the cubs a guinea fowl and was watching them fighting over it. Gopa growled most alarmingly at Jespah, Little Elsa and myself. Hearing this, Elsa instantly rushed up to see what was going on, but as soon as she had satisfied herself that nothing serious had provoked Gopa, she returned to the roof of the Land-rover.

A few minutes later, while the cubs were still eating, I went up to her; she snarled at me and spanked me twice. I retired immediately, surprised, as I did not think I had deserved such treatment. Soon afterwards Elsa jumped off the car and rubbed herself affectionately against me, obviously wishing to make up for her bad behaviour. I stroked her and she settled down beside me, keeping one paw against me. When the cubs joined us she rolled on to the other side and I ceased to exist for her.

She constantly showed how anxious she was for the cubs to be friends with us. One evening, after having gorged himself on the meat we had provided, Jespah came into the tent. He was too full to play and rolled on to his back because his bulging belly was more comfortable in that position. He looked at me plainly demanding to be patted. As he was in a docile mood I felt comparatively safe from his swiping paws and sharp claws, so I stroked his silky fur. He closed his eyes and made a sucking noise, a sure sign of contentment. Elsa, who had been watching us from the roof of the car, joined us and licked both Jespah and me, showing how glad she was to see us on such good terms.

This happy scene was abruptly ended by Gopa who sneaked up and sat on top of Elsa, with a most possessive expression which left me in no doubt that I was not wanted. So I withdrew a short distance and sketched the lions.

Fond as Elsa was of her children she never failed to discipline them when they were doing something of which she knew we disapproved, even when they were acting only in accordance with their natural instincts.

We usually kept the goats locked up inside my truck at night, but for a short time we were obliged to secure them inside a strong thorn enclosure because the truck had to go away for repairs. During this time, Jespah on one occasion besieged the boma so persistently that we were worried for the safety of the goats. All the tricks we invented to divert his attention failed to produce any effect. Then Elsa came to our aid. She pranced round her son trying to entice him away, but he paid no attention to her; then she spanked him repeatedly. He spanked back. It was amusing to watch the two outwitting each other. Finally, Jespah forgot all about the goats and followed Elsa into the tent where their dinner was waiting for them.

But when he had finished his meal Jespah, having been cheated of his fun with the goats, looked for other amusement.

He found a tin of milk which he rolled across the groundsheet of the tent until it was covered with a sticky mess. Then he took George's pillow, but the feathers tickled him, so he looked for another toy and, before I could stop him, seized a needle case which I was using and raced out into the dark with it. I was terrified that it would open under the pressure of his jaws and that he might swallow its contents, so I grabbed our supper, a roast guinea fowl, and ran after him. Luckily, the sight of the bird proved too much for him; he dropped the case, scattering the needles, pins, razor blades and scissors over the grass. We carefully collected them so they should not prove a danger to the cubs.

14
A New Year Begins

It was now time for us to go back to Isiolo and leave the cubs to a spell of wild life.

On the 3rd December I called on the District Commissioner in whose area Elsa's home lies. I wanted to give him the latest news of the cubs and to ask his advice as to how I could best use some of the royalties of *Born Free* to help to develop the Game Reserve in which she was living.

Elsa was an asset to the reserve because her story had aroused world-wide sympathy and understanding for wild life and also because part of the money I had received for her book had contributed to the sum needed to establish the new game post. On the other hand the tribesmen blamed her for the stricter supervision of poaching due to our presence. Furthermore a woman had recently been killed in Tanganyika by a tame lion and the D.C. now told me that the incident had been used to stimulate ill-feeling against Elsa. Also it was claimed that her friendship for us, by accustoming her to human beings, could make her a danger to strangers. He warned me that in the circumstances it might become necessary to remove Elsa from her home.

Four days later a rumour reached us that two tribesmen had been mauled by a lion fourteen miles from Elsa's camp. George left at once to investigate. He reached camp too late to pursue his inquiries. That evening Elsa and the cubs played happily round the tent; though they ate greedily they were in excellent condition, which was satisfactory as they had been left to themselves for seven days. As daylight broke George went to the Game Scouts' post; no one had heard of any tribesman being mauled by a lion. So he sent the Scouts to the scene of the alleged incident and returned to camp.

In order to keep the lions near to the tents he gave them a carcase which they dragged into a bush close by. They stayed there until the evening.

The day after George's hurried departure for the camp, I followed, bringing the truck as well as the Landrover. It was late when we arrived and the men were too tired to unload the truck and put the goats into it for the night. We therefore secured them in a thorn enclosure.

Although, as we had two cars, our arrival was noisy and Elsa must have heard us, she did not come to welcome me. This was the first time she had failed to do so.

After I had gone to bed I heard the cubs attacking the goats' boma. The sounds of breaking wood, growling lions and stampeding animals bleating, left no doubt as to what was happening. We rushed out but not before Elsa, Gopa and Little Elsa had each of them killed a goat. Jespah was holding one down with his paw which George was able to rescue unhurt.

It took us two hours to round up the bolting, panic-stricken survivors of the herd and secure them in the truck, while hyenas, attracted by the noise, circled round.

Elsa took her kill across the river. George who followed her saw a large crocodile making for Elsa and shot at but missed it. He spent until 2 a.m. sitting close to Elsa to see if it would reappear, but it did not. The cubs were very much upset at finding themselves and their kills separated from Elsa by the river; after half an hour of anxious miaowing they joined their mother without having started to eat the goats they had killed.

In the afternoon, the Game Scouts returned; they had not got any confirmation of the rumour that tribesmen had been mauled by lions, but they had collected plenty of evidence to show that, influenced by poachers and political agitators, the tribesmen were becoming increasingly hostile to Elsa. We realized that her life was in danger and we discussed what we should do.

We had spent six months in camp, much longer than we had originally planned, in order to protect Elsa and her cubs from poachers and by doing so had inevitably interfered with their natural life. If now we stayed on the cubs would become so tame that they would have little chance of adapting themselves in the future to the life of the bush.

Besides this, if we went on camping in the reserve we should only aggravate the antagonism of the tribesmen. Since we could not, in the circumstances, leave Elsa and the cubs alone, the only solution we could think of was to look for a new home for them and move them as soon as possible.

We had had great difficulty in finding a suitable place for Elsa's release; to find one for her and the cubs was likely to be still more difficult. We knew that by now, with their mother's help in teaching them to hunt and protecting them from natural foes, they were capable of living the life of the bush; but where would they be safe, not only from wild animals but also from man, who now proved to be their most dangerous enemy?

Leaving me in charge of the camp, George returned next morning to Isiolo hoping to find a solution to this problem.

In the afternoon I walked with Nuru to the Whuffing Rock where we had spotted Elsa. She came down at once to greet us, but when I started to climb up the saddle to join the sleeping cubs, she prevented me from doing so by sitting squarely across my path, and only after we were on our way home did she call her children. Through my field-glasses I saw Jespah and Gopa climb down, but Little Elsa remained on top like a sentry.

When it was dark the family arrived in camp and after eating their dinner, Elsa and her sons played happily in the tent until they dozed off in a close embrace. I sketched them, while Little Elsa watched us from outside the tent. In the night a lion called and for the next three days he kept close to the camp. During this time Elsa stayed in the immediate vicinity. It was only after the lion had left the neighbourhood that she ventured to take the cubs to the Big Rock and then by tea-time she returned as though to ensure an early dinner undisturbed by the possible appearance of another lion.

I usually met the family on their way to camp and was often touched by Jespah's behaviour. When Elsa and I greeted each other he didn't want to be left out, but I think he knew that I was scared of his claws, for he would place himself with his rear towards me and keep absolutely still as though to assure me that like this I would be quite safe from accidental scratches while I patted him. From then on he always adopted this attitude when he wanted to be stroked.

When reading in the tent after dark, I had lately often noticed a genet cat and heard it leaping from the ground on to the canvas roof and then taking another jump on to the tree where the weavers had their nests. Swiftly it went to the top, balancing on very thin branches, then crept towards the nests which were fastened to their tips. I focused the light of my torch on it but this did not make it give up its hunt.

It was a very young animal and it was probably due to its light weight and small size that it was able to climb along such thin branches. To reach the bird through the entrance funnel of the nest, which always faces downwards, entailed bending over the tip of the branch and was a very difficult operation. I watched the genet try many nests but, to my relief, the birds always flew out in good time. All this happened in complete silence and I was amazed that none of the escaping birds gave any warning to their neighbours that there was a murderous enemy around looking for victims.

Finally, the cat disappeared into the foliage of the trees, giving one nest a violent shaking, and soon afterwards a few feathers floated to the ground and bore witness to a tragedy. Timing the genet's activities I discovered that he caught birds at intervals of about five minutes.

By now the rains should have ended but we still had some wet days, thanks to which the bush had remained greener than it usually is in December. Perhaps it was due to this that the weavers had prolonged their breeding season.

After dark, one rainy day when the river was still in

flood, I heard hyenas chuckling from a bush just below the tent. Elsa and the cubs promptly chased them away. To judge by the growls there must have been a fight. Soon after this two lions roared upstream and Elsa replied to them. Much later I heard the cubs in front of my tent. The lions went on roaring throughout most of the night. Elsa and the cubs crossed the flooded river in the early morning; it was plain that they wished to avoid the two lions.

The 20th of December was the cubs' first birthday. It began anxiously for the river was too high to cross, so we could not discover whether Elsa was all right after the excitements of the night. I was very happy when, about teatime, the family turned up. They were wet but unharmed.

As a birthday treat I had a guinea fowl, which I cut up into four portions so that each should have a share. After gobbling these tidbits Elsa hopped on to the Landrover while the cubs tore at some meat we had prepared for them.

As all the lions were happily occupied I called to Makedde to escort me for a walk. As soon as we set out Elsa jumped off the car and followed us; then Jespah, seeing his mother disappear, stopped his meal and ran after us, and we had not gone far before I saw Gopa and Little Elsa parallel to us chasing each other through the bush.

When we came to the place where the track comes nearest to the Big Rock, the lions sat down and rolled in the sand. I waited for a little while and watched the setting sun turn the rock to a bright red; then since Elsa looked settled, I walked back, expecting the family to spend the evening on the rock. I was surprised when she followed me. She kept close so that I could help with the tsetse flies, and Jespah trotted next to us like a well-trained child. Gopa and Little Elsa took their time; they scampered about a long way behind us and we often had to stop to wait for them.

Elsa seemed to have come along just to join me in my walk; this was the first time she had done so since the cubs were born. I thought it a charming way of celebrating their birthday.

When we arrived in camp Elsa flung herself on the ground inside my tent and was joined by her sons who nuzzled and embraced their mother with their paws. I sketched them until Elsa retired to the roof of the Land-rover and the cubs started to eat their dinner. When I was sure that the cubs would not observe me I went over to Elsa and stroked her and she responded very affectionately. I wanted to thank her for having shared her children with us during their first year and having shared her anxieties during the period which is so full of dangers for any young animals. But, after some time, as though to remind me that in spite of our friendship we belonged to two different worlds, a lion suddenly started roaring and after listening intently Elsa left.

Next morning we found the spoor of a lioness upstream, but no trace of Elsa. She did not turn up that day or during the following night. On the second night we heard two lions roaring and understood why she had not come to camp. I was, therefore, astonished to see her next morning about 9 a.m. on the Whuffing Rock, roaring as hard as she could. I called to her but she paid no attention and went on roaring for an hour. To whom was she calling at this unusual time of day?

She brought her cubs in for dinner that night but when a lion started roaring she left at once, crossing the river.

Elsa and the cubs spent the night of 23rd December in camp and after breakfast when I strolled along the road to read in the sand the report on last night's visitors, she and the cubs followed me. I called to Makedde and we all walked along together for about two miles.

Jespah was particularly friendly, brushing against me, and even standing quite still while I removed a tick which was close to one of his eyes. We observed two jackals basking in the sun; on earlier walks I had seen them in the same place and they had never shown any fear at our approach. Now, although we were only some thirty yards from them, they did not move and it was only after Elsa made a short rush at them that they sneaked away, and the moment she turned back they peeped around the bushes, seemingly quite unalarmed.

We went on until we came to a rain pool where the lions had a drink. By now the sun was getting hot and it would not have surprised me had Elsa decided to spend the day in this place, but good-naturedly she turned back when we did and trotted slowly home with us.

I could not help feeling as though we were all taking our Sunday family walk. Though in fact this was the morning of Christmas Eve, and Elsa could have no knowledge of special days, by a strange coincidence she had chosen a day I felt the need to commemorate by coming for a walk with me and bringing her family with her.

When we reached the place where we had seen the jackals we found them still there and as the lions were too lazy for a game, the jackals did not even bother to get up as we passed.

Elsa and the cubs were feeling the increasing heat very much and often stopped under the shade of a tree to rest, yet when we came near the Big Rock they suddenly rushed at full speed through the bush and in a few leaps reached the top, where they settled among the boulders. I scrambled after them as best I could, but Elsa made it quite plain that I should now leave them alone. She always knew exactly how much she felt it was fitting for her to give to each of her two worlds, so I confined myself to taking some photographs of her guarding her cubs.

George arrived about tea-time with a suitcase full of mail. While we strolled about picking flowers for Christmas decorations, he told me of the inquiries he had made about finding a new home for Elsa and the cubs. He thought that the Lake Rudolf area would be the place in which the lions would be safest from human interference. He had obtained permission from the authorities to take them there if the need arose, and was soon going to reconnoitre the region to find a suitable spot.

This part of Kenya is very grim and conditions are tough there, so I felt depressed at the prospect. To make matters worse Elsa chose this moment to join us on our way home; behind her the cubs were playing happily along the road, and I could not bear to visualize them roaring on the windswept, lava-strewn desert which surrounds the lake.

When we reached camp we gave the family their supper which kept them occupied while I arranged the table for our Christmas dinner. I decorated it with flowers and tinsel ornaments and put the little silver Christmas tree I had kept from last year in the middle and a still smaller one which had just arrived from London in front of it. Then I brought out the presents for George and the boys.

Jespah watched my preparations very carefully and the moment I turned my back to get the candles he rushed up and seized a parcel which contained a shirt for George, and bounced off with it into the thicket. Gopa joined him immediately and the two of them had a wonderful time with the shirt. When at last we rescued it it was in no state to give to George.

By now it was nearly dark and I started to light the candles. That was all Jespah needed to make him decide to come and help me. I only just managed to prevent him from pulling the tablecloth, with the decorations and burning candles, on top of himself. It needed a lot of coaxing to make him keep away so that I could light the rest of the candles.

When all was ready he came up, tilted his head, looked at the glittering Christmas trees and then sat down and watched the candles burn lower and lower. As each flame went out I felt as though another happy day of our life in the camp had passed. When all the lights had gone out the darkness seemed intense and as though it were a symbol of the darkness of our future. A few yards away Elsa and her cubs rested peacefully in the grass, hardly visible in the fading light.

Afterwards George and I read our mail. It took us many hours to do so, during which our imaginations travelled across the world and brought us close to all the people who were wishing Elsa and her family and us happiness.

Mercifully it was one of the last envelopes I opened which contained an order for the removal of Elsa and her cubs from the reserve.

Elsa's Camp, 24th December, 1960

PUBLISHER'S NOTE

A month after Joy Adamson finished this book Elsa died in the bush after an illness lasting several days. A post mortem established that she died from babesia, a parasite which destroys the red blood corpuscles.

The cubs immediately became very wild and, for a few weeks, only came to the camp after dark to be fed. Then they disappeared.

Shortly afterwards the Adamsons learnt that they had been attacking goats belonging to local tribesmen and it became essential to catch them and move them to an uninhabited area. This highly difficult operation which involved trapping the cubs and transporting them 700 miles to the Serengeti National Game Park, Tanganyika, was achieved in May 1961.

APPENDIX

Record of wild lions' visits since Elsa was sired

1959
Sept. 16 Elsa's lion called from nearby during night.
      19 Elsa's lion called, Elsa disappeared upstream.

20–10th October – Isiolo.

Oct. 10 I observed that Elsa was pregnant.
     12 A lion called all night – Elsa ignored him.
     13 A lion called during morning from release place. We left Elsa there with zebra bait. Elsa kept away all day. Elsa followed a lioness spoor in opposite direction. A lion called in the evening from release place.
     14 A lion called in the early morning ten yards from camp. Elsa away.

15–30 Isiolo.

     30 Elsa acted as aunt to wild lioness. A lion called.
Nov. 1 Elsa's mate called several times close to camp.
      2 Elsa's mate called at 7 a.m. Elsa joined him – returned at 5 p.m. He called all night, Elsa walked between him and camp.
      3 Elsa's mate called at night.

4–11 Isiolo.

     12 Elsa's mate called from Big Rock, in response to siren. Elsa in camp.
     13 A lion called from Rock during night.
     15 Elsa and her lion hid in cleft of Whuffing Rock at 5 p.m. Elsa spent night in camp.

16 Elsa not in camp. I heard two lions in afternoon. Elsa returned. Her mate called from close to camp during night.

17 Elsa spent day with her lion across river. Spoor observed. She returned at 5 p.m. to camp. Her lion called during early part of night.

18 Elsa absent. Her lion came to 'kill' at camp, then crossed river. Elsa returned to camp at 5 p.m. – left after dark. Her mate called.

19 Elsa away all day.

20 Elsa called lion to join zebra 'kill.' Cleared off at 3 p.m. Her lion called from Rock.

21 Elsa called her lion to zebra 'kill,' then went off across river.

22 We found her lion's spoor in 'kitchen lugga.' Elsa went off for night.

23 Two lions called round camp. Elsa absent but heard her from Rock.

24 Elsa went off three times during night. Her lion called.

25 Elsa left 'kill' behind car for her lion.

26 Elsa's mate dragged 'kill' to target tree. Both spend day across river. Elsa returned 4.30 for food. Her lion called after dark, dragged meat we left for him away, ate it. Elsa gone by morning.

27 Elsa remained on Big Rock – called to her lion – spent all night away.

28 Elsa spent night in camp. Her lion called from far away during the night.

29 Elsa gone by morning. Returned across river at 5 p.m., walked with us to Big Rock – we heard her lion close by. Elsa remained on Rock for night.

30 Elsa in camp – left early morning.

Dec. 1 Elsa returned to camp 5 p.m. Her lion called all night. Elsa in tent.

2 Found pug marks of Elsa's lion leading to camp.

3 Elsa in camp, ate a lot, went off several times towards Rock.

4 Elsa in camp. Behaved as she did yesterday.

6 Elsa chased cow herd of buffaloes. Stayed away during night, but we heard her calling to her lion.

7   Met Elsa at Rock at 5 p.m. She returned to camp, ate a lot. Called all through the night to her lion.

8   Elsa in camp. Called all night to her lion.

9   Elsa called all night to her lion. He replied. She had gone by morning. I heard her call from far away. At 8 p.m. she returned, ate a lot, then roared full strength near river bush, while her lion roared full strength near kitchen. Both went off soon afterwards.

10 Found Elsa's spoor and that of her lion walking up to 'kitchen lugga.' Saw Elsa in the evening on Big Rock. She returned for food. Her lion called.

11 Elsa all day in camp. Her lion called during night.

12 Elsa in camp, followed us to cataract. Her lion called at night.

13 Elsa on Big Rock in morning. Her lion called from across river.

13–16 Isiolo.

16 Elsa waited for us in camp – very hungry.

17 Elsa in camp, ate a lot. She went to Big Rock but returned soon afterwards.

18 Elsa in camp – did not join us for walk.

19 Elsa near camp in morning. In afternoon joined us for walk, often sat down and finally disappeared into bush. Spent night away. Called early evening and early morning from direction of Rock.

20 Found Elsa on Big Rock in the afternoon – in labour.

Birth of cubs

21 Father lion called during night. Followed his spoor next morning coming from Hyrax Rock to camp, dragged high meat away and ate it, left via 'kitchen lugga.'

22 He called during night. Spoor found round camp but he took no meat. Later he dragged heavy water buck from Big Rock half a mile across bush. In afternoon found his spoor leading up lugga to water buck. Met him under bush, he growled when Toto crouched – then went off.

23 Found father's drinking spoor along river.

24 Father came to camp and ate high meat. During night he came again and tore at hanging meat.

25 Elsa returned after five days' absence. Father crossed river to other side.

26 Father called from across river, far away.

27–31 Isiolo.

1960
Jan. 3 A lion called during night – took goat-head from camp. Two lions' spoor – showed they had followed road towards Big Rock. A lion called during night.

5 A lion called during night at some distance.

28 Father called from across river. Elsa absent.

29 Elsa roared loudly about 5 p.m. A lion called during night. Elsa absent.

30 A lion called during night. Elsa absent.

31 Elsa arrived at 9 a.m. from across river – roared as loud as she could from the middle of the stream and returned. She came back at 4 p.m. ate a lot and left at 6 p.m. A lion called during the night.

Feb. 1 Elsa absent all day. During night she came to camp, jumped at my truck to get a goat, stopped when I called 'No.' Father called from across river.

2 Elsa introduced her cubs to me.

8 Father called in early dawn.

10–14 Isiolo.

21 Father called during night from 'kitchen lugga.'

22 Leopard barked and lion called all night. Elsa absent.

23 Father called during night from Rock.

29 Father called evening and early morning from nearby.

Mar. 12 Makedde observed lion pug marks in 'kitchen lugga.' Father called while Elsa remained in camp.

13 Father called during night from far off. Elsa stayed in camp.

18 A lion called upstream during night.

26 Father stole goat remains.

28 Father called from close by. George nearly walked into him hiding in bush four yards away from tent. Terrific growl, quick retreat by George.

April 1 Father called during night.

5 Father called during night.

7 Father called during night.

10 Found father's spoor along road.

20 Father called soon after dark from close by. Elsa and cubs went off. He called again from farther away.

21 Father called from close by.

24–28 Isiolo.

May 4 Father heard close by.

5 Found father's spoor along road.

6 Elsa behaved oddly, was silent as if her mate might be close by. She concealed the cubs.

7 Found father's spoor in 'kitchen lugga' leading to Big Rock.

8 Found father's spoor close to camp. Elsa roared from Land-rover after dark.

9 Found father's spoor on road. Elsa hidden for two days near camp. Found father's fresh spoor superimposed on George's car tracks. Elsa hid below camp with meat and her cubs.

10–18 Isiolo.

18 Elsa hid carcase near doam palm, kept away all evening. Father called from Rock. Elsa appeared at 3 a.m. got on to George's bed.

22–31 Isiolo.

June 4 Father called in early evening. Elsa and cubs hid at buffalo path.

6 Found father's spoor near kitchen where he stole a boiled guinea fowl.

7–16 Isiolo.

21–1st July Isiolo.

July   3   During our absence the camp was burned. A lion called from the Big Rock direction. Elsa called from across river after having had a fight. Cubs lost. Spoor of two lions all around Big Rock.

      4   At night heard lions from Rock. Found lion and lioness spoor at water buck bay. In evening a leopard coughed and a lion growled close by. We also heard a hyena.

      8   Father, a leopard and a hyena called during night. Elsa in camp.

      9   In the evening a lion called. Elsa rushed off towards 'kitchen lugga', leaving cubs behind. After an hour she returned, crossed the river. Leopard and lion noises all night. Hyena stole goat remains.

   10   Elsa absent all day. Father called during night. I found his spoor in 'kitchen lugga' with those of a lioness.

   15   Lion growled from upstream. Jespah missing. Fight between lions. Elsa badly mauled.

   16   Found lion spoor upstream on sandbank – mixed with Elsa's. That night, while Elsa was in camp, a lion called from far away.

   17   Fire below kitchen during night. Elsa across river. Two lions ate goat remains near tent at dawn, when they crossed the river, growling loudly.

   19   Two lions came to camp during night. One growled near goat truck, then went off 'whuffing' to Big Rock. Elsa absent.

   22   Lions called at night. Elsa absent.

   23   Lions called at night. Elsa absent.

   24   Lions called at night. Elsa absent.

   25   Lions called at night. Elsa absent.

   26   Tracked lion's spoor near baobab. Could have been those of Elsa and her cubs. Found single lion spoor upstream at drinking place.

   27   Tracked lion, lioness and cub spoor five miles upstream crossing river, saw lion thirty yards away, heard him roaring.

   29   Found single lion and two lions' spoor five miles upstream.

   30   During night a lion called very close, another very far off replied. Found lion's spoor coming from Rock to camp.

31 Elsa returned without cubs after seventeen days' absence. Off after half an hour.

Aug. 1 Elsa brought cubs to camp. Boys tracked their spoor five miles downstream, where they found them mixed with spoor of two lions.

7 Elsa was alert, then left. Two or three lions came near kitchen – terrific growling. One lion crossed river.

8 Found Elsa with mauled paw and cubs beyond Border Rock. Took them home. Elsa would not eat. Soon left with cubs. She sprayed her jets along bushes. At night called from Big Rock.

9 George found two-day-old spoors of two lions running fast in elephant lugga.

11 A lion called close by after dark, moved upstream. Elsa left camp before midnight – returned moaning, then left.

12 Found one lion spoor at end of studio lugga, and another at rhino carcase on opposite side of the river.

13 Lion called from across river.

18 Two lions approached from upstream. Elsa when challenged left cubs behind. Returned later – cubs had by then disappeared. Two lions roared near kitchen. Cubs returned – Elsa took them away.

22 Father called.

24 Two lions approached from upstream. Elsa fought the fierce lioness.

25 Found Elsa badly mauled down river.

27 Found lioness's spoor coming from upstream to camp and then going back.

Sept. 2 Two lions called from upstream. Elsa took cubs to studio and went off to tackle these lions. On her return the cubs had disappeared. Elsa went searching for them. Found lioness's spoor parallel to Elsa's spoor along road.

5–8 Isiolo.

9 Two lions called upstream. Elsa replied from across river.

12 A lion called from upstream.

13   Elsa lay in tent, very ill. Cubs were missing – returned at dusk. All crossed river at midnight. Fierce lioness called soon after from Rock.

14   George saw fierce lioness on Rock. Elsa away all day. Fierce lioness called during the night.

15   Found Elsa and cubs beyond Border Rock. She was badly mauled. Brought them home.

21   During early part of night a lion called upriver. Elsa went off at midnight. Lion called again. Elsa returned with cubs early morning at breakfast time.

22   George tracked single lion's spoor in 'kitchen lugga' leading to pig bay.

Oct.   7   During night a lion called from Big Rock.

8   A lion called from Hyrax Rock.

12   A lion called from far away.

16   Heard fierce lioness and her mate calling from Big Rock at night.

17   Heard fierce lioness at dawn. George saw her soon after on Big Rock. He stalked her to within four hundred yards – then she disappeared.

20   While Elsa and cubs were in camp, two lions approached from upstream. Elsa and Jespah left instantly, other cubs followed them across river only after two wild lions had roared from very close. Two lions roared all night from Rock.

21   From the early morning a lion called from the Rock. At dusk Makedde and I saw the fierce lioness on top of the Big Rock. She watched us. Elsa absent. All through the night the two lions called from the Big Rock.

22   Makedde tracked two lions' spoor leading upriver.

29   A lion called from across the river. Elsa went off downstream.

30   A lion called from far away. Elsa went off across river.

31   Found single lioness's spoor going up the road to the rain pool, then back to 'kitchen lugga' and off upstream.

Nov. 2 A leopard coughed from across the river. Elsa left the cubs behind and chased him off. Then they all crossed the river. Later a lion roared close to camp.

3 Found lion spoor at 'kitchen lugga' and Big Rock.

13–21 Isiolo.

Dec. 11 I heard a lion calling from upstream all night. Makedde heard two.

12 Makedde told me he had heard a lion in 'kitchen lugga.'

14 Found fresh lion spoor along road. Elsa and cubs kept close to camp during last two days. She never left.

18 A lion called during night from across river.

19 Two lions approached from upstream; lots of 'whuffing.' Elsa joined in their noises. 'Whuffing' went on all night. Elsa and cubs crossed flooded river in the early morning. Elsa defended goat carcase from hyena.

20 A lion called during night.

21 Found single lioness spoor in 'kitchen lugga.' A lion roared all night. No Elsa.

22 Two lions roared during the night, upstream. Found Elsa at 9 a.m. on Whuffing Rock – roaring. I called her from release track. She ignored me – went on roaring until 10 a.m. Could not find her all day. At 8.30 p.m. she came with cubs to camp. One lion called all night upstream.

26 A lion called during night from upriver. Elsa took cubs across river.

27 Found fresh lion spoor at rain pool along road in afternoon.

28 Father called close by in early morning. Elsa away all night. Found fresh lion spoor on road near rain pool.

29 No Elsa, but heard her roaring from Rock.

30 Found spoors near Rock, crossing road towards 'kitchen lugga.' Elsa away from 28th to 30th.